Technologies of Being
in Martin Heidegger

Technologies of Being in Martin Heidegger attempts to deepen the dialogue between philosophy of education and philosophy of technology while engaging with the thought of Heidegger, Jacques Derrida and Bernard Stiegler. Through a critical reading of Heidegger's central notion of *nearness*, this book argues that thinking is intricately conditioned by technologically produced images, which are themselves interacting with the imagination's schematising power.

The book further discusses how certain metaphorical synthesising processes, which are currently industrialised in the form of social networking sites and search engines, discretise human behaviour and reorganise it in ways that often marginalise human interpretation and redefine nearness. Finally, it suggests how we might reconceptualise technology and education as processes of human individuation.

Technologies of Being in Martin Heidegger will be of great interest to scholars in the fields of philosophy of education, philosophy of technology, literary studies, cognitive linguistics and cognitive neuroscience.

Anna Kouppanou is a postdoctoral research fellow at the University of Cyprus.

Routledge International Studies in the Philosophy of Education

For more titles in the series, please visit www.routledge.com/Routledge-International-Studies-in-the-Philosophy-of-Education/book-series/SE0237

Technologies of Being in Martin Heidegger

Nearness, Metaphor and the Question of Education in Digital Times

Anna Kouppanou

Routledge
Taylor & Francis Group

LONDON AND NEW YORK

First published 2018 by Routledge

2 Park Square, Milton Park, Abingdon, Oxfordshire OX14 4RN
52 Vanderbilt Avenue, New York, NY 10017

Routledge is an imprint of the Taylor & Francis Group, an informa business

First issued in paperback 2019

British Library Cataloguing-in-Publication Data
A catalogue record for this book is available from the British Library

Library of Congress Cataloging-in-Publication Data
A catalog record for this book has been requested

ISBN: 978-1-138-22068-3 (hbk)
ISBN: 978-0-367-23279-5 (pbk)

Typeset in Galliard
by Apex CoVantage, LLC

To Lukas

Contents

Credits list

Foreword

The work of the imagination in education – that is, in our coming into the world and becoming human – rarely receives the attention that is its due. It is not, of course, that imagination is not a research topic or that the concept is never submitted to analysis (though this happens less than one might expect). It is not that it is absent from policy or from the everyday discourse of practising teachers. One child "lacks imagination" while another "shows remarkable imagination". Imagination is the stuff of "creativity". It is exercised in the art class or in creative writing, sometimes in science too. Maybe, more moralistically, it extends into a child's ability to emphasise with others, to see things from a point of view other than their own, to walk for a while in another person's shoes.

Now all this is important, no doubt, but it circumscribes the notion with disastrous consequences – disastrous, that is, for our understanding of what it is to come into the human world, of what that is meant by "world". In education this disastrous underestimation must be viewed with a kind of irony. After all, there where it is precisely the notion of coming into the human world that is of pivotal importance, we find it so typically distorted. There where this matters so much, we find a vacuum that developmental and positive psychologies rush to fill. The irony here is all the more painful when it is acknowledged that there *is* a robust account of imagination that has been established, now for most of a century, and whose influence within philosophy has been profound, and this, of course, is centred in the philosophy of Martin Heidegger. Half the philosophical world may have remained deaf to Heidegger's thought, but his influence and importance have become more difficult to ignore. For sure, Heidegger's thought is challenging, even impenetrable in parts, and it has never freed itself from his notorious political involvements, from misjudgements and misguided sympathies whose implications will continue to reverberate through the reception of his work. But his crucial contributions to and extensions of phenomenology laid the way for some of the most trenchant critiques of the world we live in, while at the same time articulating a conception of being-in-the-world whose importance for education is not only immense but, for the most part, still to be received.

The idea of the imagination is central to this because imagination is crucial to the very workings of human thought and, hence, to the idea of world as this emerges in a mutual dependence with that thought. Crucial to human thought, unlike that of animals, is a play of the imagination realised in forms of indirectness. Things are understood in terms of their possibility. The cup is for drinking from, the table for leaning on and the keyboard for composing a foreword to a book. These things make possible those activities, and they are not intelligible as what they are without reference to such possibilities: to such possibilities and to others, for it is also true that the cup may be adapted for other purposes, perhaps to hold an assortment of pens and pencils, perhaps to pot a plant. Such possibilities are apparent because we can see one thing as another, see the empty cup in relation to these different possibilities. And there always will be these different possibilities. We take one thing *as* something, and it is always the case, as we mostly know, that other possibilities are there too. Our notions of following a rule, of participating in a practice, of knowing how to go on, in Wittgenstein's phrase, depend precisely on this. 'Use the cup properly' makes sense only insofar as we can conceive of using it 'improperly', and no sense of propriety will secure this sense and usage from those other possibilities: they are the condition for the very notions of propriety or correctness, which is more or less to say for meaning anything at all – meaning, that is, in the ordinary way that meaning figures in a human life and figures in what is meant by world. What could education be without this?

Crucial to the *as-structure* of thought which was sketched here, are qualities of openness of possibility, as we have seen, but also of indirectness: to see something as something else is to make a connection, with something other than our starting-point, typically with something not here or not now; and paradoxically it is in relation to such possibilities that our starting-point emerges. Supreme elements in Heidegger's depiction of this are, first of all, tools and what we do with them, and, second, language and what it does with us. The mediations these provide are not an addition to human experience, an extension of our powers, let us imagine; they are, on the contrary, constitutive of that experience. They enable us to project and construct a home, a city, a world. In this sense, our city *is* a city of words. And they make us what we are. The indirectness through which they work is always a matter of making connections, in which things are brought near. What is close – in fact, this whole dimension of what is near and what is far – is quite unlike the distances measured in physics, and it is more fundamental to how sense is made of things. It is a matter of significance, without which the world and human beings cannot be.

To make connections is to carry across, such that things are understood in terms of their possible relations. Carrying across, transfer, μεταφορά . . . The lines of connection, etymological and practical, are not to be resisted, linking as they do what words do and what tools do and returning the abstract-theoretical term *metaphor* to its everyday practical origins in Greek. This is not to say that words are just tools, far from it, for language, Heidegger tells us, is the House of Being, which is to say that it is in language that we can be and in language that the world is. But then, paradoxically again, nor are tools just tools (say, instruments that we simply 'use' for a particular end), for they, too, open up the world in constantly new ways.

The Greek sensibility that Anna Kouppanou brings to these matters in *Technologies of Being in Martin Heidegger* enables not only a richly rewarding interpretation of central aspects of Heidegger's work but also a searching examination of aspects of education crucial to our understanding of it today – that is, education in digital times. For in digital technology we are confronted not merely with a new technological device but also the realisation of new possibilities of language: these are at once openings onto new possibilities of life and world, and threats to contemporary societies and civilisations. They are threats especially in terms of the characteristic manner in which they 'enframe' our lives, our circumstances and the possibilities of thought. They are threats to education in the multiple ways that are familiar to teachers – from the intrusion of mobile phones into the classroom to the ubiquitous presence of social media and to the new prominence of virtual realities, albeit that all these things might enhance education if taken up in appropriate ways.

It is an outstanding feature of Kouppanou's book that, by drawing out the profound importance of metaphor in Heidegger's work, while at the same time retrieving the notion from its contemporary, non-Greek, intellectualised sense, it takes the reader through a particular reading of Heidegger to a point where practical questions of digital technology and social media can be addressed in relation to their significance for education. The book enables us better to understand the world we live in and who we are, now that we are living during these times. This is far, however, from a dystopian vision, in which digitised new technology is demonised. Kouppanou is at pains to show that technology has never been isolated from other forms of human dispersion – through embodiment, language and materiality. Education is difficult but not impossible. The immense power of the response to technology that is found in the work of Karl Marx is motivated in part by his confrontation by the mostly one-tracked emergence of technology at that time – the technology of the factory production line. The encroachments of that technology on other aspects of life – including, for example, the later influence of Taylorism on education – can readily be seen. But the complexity of the processes in which contemporary technology participates may prevent it from so easily becoming dominant. The diversity of new technology's interaction with other aspects of human experience may open the way to a kind of critical thinking that resists domination by any unlinear forms of enframing. The imagination will be exercised in new ways. And the commitment to *Bildung* that technology has in the past threatened may be recovered as thought is, thus, carried across. This more affirmative note, with which Kouppanou concludes, provides the way to think more realistically and more pragmatically about the real possibilities and challenges that education now faces.

Paul Standish
UCL Institute of Education

Introduction

We are now closer than ever!

The word *closer* exudes emotion. As a whole, the sentence connotes content-ment, accomplishment or even surprise, but it is of course taken out of context, having at the same time and in some inexplicable way already allowed the forma-tion of an expectation in us. The sentence is ready to be projected, or perhaps it has already been projected onto some memory or phrase, a notion or a senti-ment, leading us to think, experience or feel a specific something or a specific type of closeness. However, the word 'closer' is not the only source of ambivalence here – the other peculiar word positioned around the middle of the sentence, the word 'now', urges us to wonder if the sentence suggests a position in *space* or a point in *time*. Still, is there a position that is just in space or a point that is just in time? What's more, is there even space and time without our being *at some point*? When we use the word *close*, we may indeed refer to time, to an emotion or even to a condition in which we find ourselves thrown. We say that *we are close to somebody*, and we almost never mean near to each other, touching one another. We speak in metaphor. But is not movement, any movement for that matter, a movement of closeness, and does the word *metaphor* not connote, in ancient and modern Greek, a movement in space?

In any case, to be close is to have a relationship with that which exists in prox-imity but also with that which remains far. Presence and absence are part of the game of nearness; that is, if nearness does play games. For the German philoso-pher, Martin Heidegger, to be a human being – that is to say, to be a being in time that interprets time as its way of being, means to be a being of and in *near-ness*. The human being, Heidegger claims, is defined by its propensity to near-ness. Language and technology play an important role in the drama of nearness: you can be close to a thing – to a tool that is in your very hand, like my hand is now close to the keyboard, without having the slightest contemplation about this implement or the fact that you are close to it. The tool does not matter to me, even though the task that I am pursuing, matters a great deal and is in fact possible to take place through this tool that I am using. You can also be close to a person who is far away, in point of fact, a person who might not even *be* at all, but *close-ness* remains a possibility. 'I am close to my brother or sister', you say. You can even talk to them from afar, you can think of them, you can actually be there for them through technology and language – you can call them, write to them, send them a card and thus experience closeness even when you are far away. Language and technology allow you to show concern and care. They matter to you through the matters that they allow to take place.

Of course, one could object, or in fact simply state, that the phrase at the beginning of this book is beyond doubt metaphorical: to be close is to be metaphorically close; being close is being in metaphor. This argument would, however, only complicate things since being in metaphor can mean many different things to many different people. We have to admit that there is here an interpretive freedom, but, after all, to be in metaphor is always already to be in *interpretation*. The reason that the starting phrase of this book – taken completely out of context, indeed, having no context at all – can have meaning for us is because we are interpretative beings. Our being in the world is, according to Heidegger, to have always already interpreted this world *as* a certain world. To be close is to be in interpretation; to see things as certain things. To be in time is to be in meaning. For Heidegger the human being is always already situated in a hermeneutic-existential circle that forms and transforms a reality that does not exist prior to its interpretation. This process is the way the world comes close: essentially through *connection* and *schematisation*. So, to be close to anything is to use metaphor.

However, language, as Heidegger (2008) tells us, is not something used. It is rather something in which we unfold *as* the beings that we are. This *as-structure* is what nearness is really about. It refers to the possibility of the emergence of distinct beings and assumes many forms in Heidegger's work. It is interpretation, time, the unfolding of beings and *Being*, the revealing of truth and the poetics of the world. Above all, I argue here, through a deconstructive reading of Heidegger, that nearness is imagination. It is a process of formation and a way of knowing that unfolds through the transformation of the very things it brings near. Therefore, in order to understand imagination, and what's more imagination in Heidegger, we need to shed light on his central notion of nearness and on a chain of notions that make up this 'new economy of relating' (Ziarek, 2000, p. 139).

The tracing of nearness would appear to be embracing here the biographical approach of Heidegger's work that refers to early and to later Heidegger (Luchte, 2008; Pattison, 2000), but it would do so with a critical disposition. By that, I mean that I present Heidegger's work in these stages (early, middle, later), which are not always kept in firm distinction from each other, in order to study nearness in its continual and untiring metamorphosis. In other words, the adoption of the biographical approach does not imply the adoption of the pre-given interpretations of Heidegger's work that neatly differentiate between the early phenomenological-hermeneutical and existentialist approach, the unfortunate philosophical misreadings of the middle period, and the post-subjectivist, non-phenomenological and poetical approach thought to be embraced in the later period. The chronological distinctions are used here only provisionally, and perhaps instrumentally, in order to show that despite the supposed fragmentation of Heidegger's thought, there is a stability of concerns. In this respect, I see what I am doing as congruent with Iain Thomson's approach (2005), which pays heed to Heidegger's persistent effort to de(con)struct Western metaphysical structures, while it is also in line with Heidegger's own reading of his work, as expressed in the preface to William J. Richardson's (2003) book, *Heidegger:* Through *Phenomenology* to *Thought*. In the preface, Heidegger (1962) informs Richardson that:

> The distinction you make between Heidegger I and II is justified only on the condition that this is kept constantly in mind: only by way of what [Heidegger] I has thought does one gain access to what is to-be-thought by [Heidegger] II. But the thought of [Heidegger] I becomes possible only if it is contained in [Heidegger] II.
>
> (cited in Richardson, 2003, p. xxii)

Heidegger's reflection on his own work also reveals a basic premise of his philosophy and of his way of doing philosophy: chronology and causality do not suffice to explain the emergence of meaning and certainly not when these notions remain themselves unquestioned.

Questioning *Being* is precisely Heidegger's question. This book attempts to take into consideration Heidegger's expressed concerns but also the neglected continuities of his work in order to allow his way of doing philosophy to bring to the fore that which his philosophy addresses. A central concern of this thought is language's and technology's participation in the giveness of beings. Heidegger's descriptions of language and technology, however, sometimes appear to underscore these processes' synergetic role and at other times appear to highlight their antagonistic rapport for the constitution of Being. This conflict is magnified in Heidegger's later writings in which modern technology is depicted not merely as prohibiting human beings to experience nearness but also as fabricating what the philosopher calls *distancelessness*. In his lecture *The Thing*, Heidegger (1975) expressly addresses this matter, asserting that '[a]ll distances in time and space are shrinking', and adding that 'the frantic abolition of all distances brings no nearness' (p. 165). It rather produces an experience that is quite representative of technology's impact, and that Heidegger calls 'distancelessness' (p. 166). Expanding on this topic, in his well-known essay *The Question Concerning Technology*, Heidegger (1977) describes modern technology's essence in terms of *Gestell* or 'Enframing' as it has been rendered in most English translations (p. 15). However, according to William Lovitt (1977), who is the translator of the aforementioned essay, the German etymology of *Gestell* also bears the connotations of setting in place, order, command, entrapment, disguise, representation, production and presence, while Véronique M. Fóti (1985) points out that the word *Gestell* connotes the formation of the human being and whatever relates to it. For Heidegger, *Gestell*, as the essence of technology, forms the world and forces everything to present itself as resource or stock. This *technologisation* of life corrodes everything, the human being and its way of being in the world included. It is no surprise then that the Heideggerian critique of *Gestell* has been reiterated as a form of educational critique in order to discuss technology's poisonous effects on learning, which are usually contrasted to the curative forms of connectedness to the world – like craftsmanship, art and poetry. What this book aims to do, however, is to look into the equivocal possibility that lies at the heart of all of these types of nearness and to investigate their *schematising forming* potentiality that is inherent in both technology as *Gestell* and education as *Bildung*. In more detail, the book attends the following trail:

> The *first chapter* begins from the end. Educational critique ensuing from Heidegger's later and explicit statements about the essence of modern technology is discussed, asserting that this type of critique is useful but not sufficient. Through a close reading of Heidegger's essay *Plato's Doctrine of Truth* (2009), the chapter begins by sketching deeper assumptions of Heidegger's philosophy of education, as it interweaves essential bonds between truth and education and differentiates between truth as *alētheia* (a type of revealing and movement of nearness) and truth as *orthotis* (a type of correctness and correspondence between thinking and the world). This discussion allows us to begin understanding a web of terms, concerning image (*Bild*), imagination (*Einbildungskraft*), education (*Bildung*), and metaphor, that instantiates *Heidegger's philosophy of nearness*. This chain of notions, connoting a process of formation according to image, reveals Heidegger's focal distinction between originary/poetic image that brings true nearness and technological/representational/representative image that negates the possibility of nearness entirely. A preliminary investigation of this dichotomy suggests that it is not materiality, exteriority or repeatability that distinguishes one type of image from the other – as Derrida's critique puts forward – and, that what is at stake is not the difference between presence and representation. *In contrast, what is at stake is the relationship between types of images – relating either to language or to technology – and the kind of formation, synthesis and thinking processes that these images allow to emerge.*
>
> The *second chapter* discusses the human being's hermeneutic nature in order to indicate the ways that this nature is constituted and conditioned. Throughout his career, Heidegger was concerned with the ways the human being and beings in general emerge *as* the beings that they are and perceive beings *as* beings. This *as-structure* that lets things be as they are (*Seinlassen*) while receiving them remains a constant concern in Heidegger's work, denoting not only the receptivity but also the interpretive and transformative spontaneity of human perception. The play between passivity and creativity is mirrored in the ancient Greek word *phantasia* (imagination) which, in contrast, to its Latin counterpart, connotes the way things appear (*phainetai*) as they are. This aspect of imagination, along with the re-inscription of Kantian imagination, becomes the cornerstone of Heidegger's phenomenological method and of his philosophy of time. For Heidegger, imagination and time constitute a type of *synthesis. However, a problem arises concerning the deeper nature of this synthesis, when Heidegger*

gives priority to language's apophantical revealing (αποφαίνεσθαι) over technological synthesis. In order to examine this assertion, we need to investigate the workings of the forming power of the imagination and the ways language and technology possibly condition it.

The ***third chapter*** focuses more explicitly on the structure of the imagination so as to assert that Kantian imagination, as appropriated by Heidegger, is affected by exterior images during a process of transfer of the inside onto the outside. In other words, the 'interior' cognition relies on 'external' support in order to be formed and, in consequence, to form what is to be perceived. In this respect, the interior is not prior to the exterior. In fact, such dualism can no longer be sustained. This realisation makes *metaphoricity* – that is to say, the transfer of the interior onto supplements (technologically or linguistically produced images), extremely important. *This new take on metaphoricity relies on the Greek connotations of the word* metaphora *(μεταφορά) and suggests transfer, movement and even transportation. It is a process of connectedness: a thing is synthesised in this movement of coming close.*

As we move to the ***fourth chapter***, we look directly at the exterior technological substratum that co-constitutes the human being's, or according to Heidegger's terminology, *Dasein*'s environment. A tool is, for Heidegger, something ready-to-hand, allowing *Dasein* to pursue different ends, to approach its possible selves and to experience nearness. Heidegger, however, understands this type of nearness as inauthentic, especially when it is set against nearness towards the irreversible end of death. However, the deconstructive critiques of Jacques Derrida and of Bernard Stiegler point out that these two types of temporalities are inadvertently entangled. *Dasein*'s *already-there* is always already constituted by technological exterior support containing past memories and forming *Dasein*'s future. *Dasein* appropriates these memories during its own process of becoming or of individuation. For this reason, Stiegler calls technologies, especially those containing memories (*mnemes*), *mnemotechnologies*. These technologies are defined by their finitude: they cannot contain all memories. Therefore, during their design, a certain selection process needs to take place that ultimately conditions their use and their users. In this respect, selectivity defines the becoming of human beings and is, according to Stiegler, inherent in both linguistic and technological artefacts. The emphasis on the mnemonic nature of language and technology allows Stiegler to claim that technology is above all memory, marginalising in this way all of the other aspects of technology's schematising power. *The notion of metaphoricity, in contrast, allows for a fuller depiction of the nature of technology and the kind of nearness it brings to the fore. This chapter looks into these processes in order to prepare the investigation of technologies as metaphoric machines.*

The ***fifth chapter*** investigates nearness as it appears in the middle phase of Heidegger's thought in order to show that the exclusion of technology from the analysis of spatial notions, like *nearness, homeland* and the *polis*, results in theoretical misinterpretations and political mistakes. In fact, Heidegger's exclusion of technology from the theorisation of the social milieu turns his political thought into a theory of exclusion. The belief that there can be a type of *polis*, which is not mediated by technology but is instead rooted in a supposedly non-technologically mediated Greek beginning, leads Heidegger to align his views with the Nazi political vision. *In order to sketch this possibility, Heidegger misuses phenomenology, whilst his analysis focuses on the notion of the home', which is understood more as a spiritual connection between a land and its people and less as an everyday mundane experience that illuminates what it means to be an embodied being interacting with other beings, human or not.*

In the **sixth chapter**, I turn to Heidegger's later work and explore further his analysis of language and metaphor, and metaphor's connection to nearness. In consequence, I illuminate how metaphor relies on analogy and the copula of the verb *is*. This takes us back to the question of Being and to the question of the ontological difference. Being can *be* in analogy to beings and beings can *be* in analogy to Being. This kind of analogical and metaphorical synthesis resembles the Kantian imagination and the as-structure. Analogy is not, however, always present in metaphoricity since the domains coming together in the metaphorical movement are not always strictly demarcated. In order to examine metaphoricity more carefully, the chapter focuses on Heidegger's most important metaphor, namely, *the one that presents language as the house of Being*, in order to deepen the reconceptualisation of metaphor, as a conceptual pragmatic and material schematising process of nearness, which is related to Heidegger's originary and poetic image. *Contrary to Heidegger, however, the chapter argues that both language and technology instantiate the metaphorical process as originary image.*

In the **seventh chapter**, I summarise this new notion of metaphoricity and examine current digital technologies through Martin Heidegger's existential analytic, Bernard Stiegler's critique of modern technology, and my own emphasis on nearness and image. In order to investigate the reality of new digital technologies more closely, I look at Facebook as a predominant technology of nearness of our time and analyse its metaphorising and schematising tools. *The analysis shows that Facebook's features constitute automated metaphoric schemas that form time as a certain time and nearness as a certain nearness of immediacy and nowness.*

Finally, the **eighth chapter** revisits questions specifically concerned with educational matters. It looks into various dominant perspectives in educational practices and educational discourses that appear to be based on the instrumental theorisation of technology and thus equivocally assume that technologies are neutral means with no effect on learning aims and activities and also indispensable tools for procuring and enhancing learning. These approaches are then contrasted to the metaphoricity position proposed in this book, which takes into consideration the role of technology for the constitution of the individual and society, and suggests a twofold methodology that looks into the characteristics of technologies and at the same time into the features of synthesis in general. This synthesis is embodied, conceptual, material, technological and languaging and thus cannot simply be dominated either by technology or by the human being. This perspective is also able to show how technology and education are modes of formation of the human being and at times competitive modes of *Bildung*.

References

Fóti, V. M. (1985). Representation and the Image: Between Heidegger, Derrida, and Plato. *Man and World, 18,* 65–78.

Heidegger, M. (1975). The Thing (A. Hofstadter, Trans.). In A. Hofstadter (Ed.), *Poetry, Language, Thought* (pp. 165–186). New York: Harper and Row.

Heidegger, M. (1977). *The Question Concerning Technology and Other Essays* (W. Lovitt, Trans.). New York: Harper and Row.

Heidegger, M. (2008). *Being and Time* (J. Macquarrie & E. Robinson, Trans.). Oxford: Blackwell.

Heidegger, M. (2009). *Pathmarks.* In W. McNeill (Ed.). Cambridge: Cambridge University Press.

Luchte, J. (2008). *Heidegger's Early Philosophy: The Phenomenology of Ecstatic Temporality.* London and New York: Continuum.

Pattison, G. (2000). *Routledge Philosophy Guidebook to the Later Heidegger.* London and New York: Routledge.

Richardson, W. J. (2003). *Heidegger: Through Phenomenology to Thought.* New York: Fordham University Press.

Thomson, I. D. (2005). *Heidegger on Ontotheology: Technology and the Politics of Education.* New York: Cambridge University Press.

Ziarek, K. (2000). Proximities: Irigaray and Heidegger on Difference. *Continental Philosophy Review, 33*(2), 133–158.

1 Heidegger and education

What can technology tell us about education?

1. Introduction

Heidegger's monumental *Being and Time* (1927/2008) offers an existentialist or even constructivist account of being that affirms the human being's co-constitution of self and world through its various interactions with language and technology. According to Andrew Feenberg (2006), however, Heidegger's philosophy of technology, especially as expressed in *The Question Concerning Technology*, constitutes a form of 'essentialism'. Feenberg's assessment is not difficult to understand, since, in the aforementioned essay, Heidegger (1977b) argues that modern technology has an essence – not in terms of having its nature predetermined – but in terms of being able to determine the nature of everything else. Modern technology, according to Heidegger, has a specific way of revealing things, and it does so decisively and programmatically. As such, technology's essence, consisting in 'nothing technological', is a type of *Gestell* or Enframing that conditions and forms the way we think about the world or even the way the world comes near *as* a certain world (p. 4). In this respect, Enframing is not a choice we can make or a mind-set from which we can escape but a condition imposed on us as the only reality we can possibly come to know. Heidegger in fact claims that Enframing transforms the human being and all other beings into resources, stock, and distance-less beings, asserting that:

> Everywhere everything is ordered to stand by, to be immediately on hand, indeed to stand there just so that it may be on call for a further ordering. Whatever is ordered about in this way has its own standing. We call it the standing-reserve [*Bestand*]. The word expresses here something more, and something more essential, than mere "stock." [. . .] Whatever stands by in the sense of standing-reserve no longer stands over against us as object.
>
> (p. 17)

Throughout his prolific career, Heidegger attempts to break the metaphysical anthropocentric constraints that conceptualise the human being as the controlling centre of the world. To this latter paradigm, Heidegger juxtaposes ways of knowing, which are closer to *technē*, craftsmanship, art and *poiēsis* (poetry), and that conceptualise the human being as one force, among many others, contributing to the phenomenon of world. This is because, for Heidegger (1977b), *poiēsis* is a type of connectedness that is greater than a specific domain of human activity, and he therefore translates the following passage from Plato's *Symposium* (205b): '*ἡ γάρ τοι ἐκ τοῦ μὴ ὄντος εἰς τὸ ὂν ἰόντι ὁτῳοῦν αἰτία πᾶσά ἐστι*', to mean that '[e]very occasion for whatever passes over and goes forward into presencing from that which is not presencing is *poiesis*' (p. 10). The use of the term *presencing* is quite important, clearly depicting Heidegger's intention to move away from the metaphysical static understanding of *Being* as presence and into a procedural one. This redefinition, as we will see later, allows Heidegger to locate both similarities and differences between the workings of poetry and technology: for Heidegger, poetry allows things to be as they are, whereas technology forces them to appear in predetermined ways. The question of Being is therefore transformed into one of presencing, difference, production, and *Bildung* – that is to say, education as a process of formation. In order for this to become clear, we need to look next into the ways Heidegger's philosophy of technology morphs into a philosophy of education.

2. Enframing and educational critique

Educational theory has itself translated the discourse of Enframing mostly by detecting modern technology's impact as instantiated in education' obsession with improvement, assessment, efficiency and measurement. The assumption here is roughly the following: if Enframing tends to turn the complexity of the world into measurable units of energy, it equally transforms the complexity of human beings into measurable sets of skills, pieces of information and criteria of assessment (Fitzsimons, 2002; Peim & Flint, 2009). This critical approach emphasises Enframing's versatile presence in education in the form of influences on policy formation, of distinct technologies forcefully incorporated in the teaching process and of world views that comprehend everything, education included, as an instrument for further development and increased productivity.

Paul Standish (1997) offers such an Heideggerian account of the 'technologization' of education, arguing that the field of further education[1] is formed by 'the new managerialism with its vocabulary of efficiency and effectiveness, choice and markets', locating an ally in 'a certain legacy of progressivism', 'child-centred primary education' and even 'the pedagogy of oppression derived from Paulo Freire' (p. 440). The theoretical justification for such an amalgam of an educational philosophy is easily discerned: an efficiency-oriented plan with pregiven standards should be set in place so as to ensure that individual needs are being met. This kind of plan, however, presupposes that needs, performances, and learning itself is measured in the same way for each individual being. In this light, human beings are translated into numbers, whilst actual individual differences become inessential. Students get to experience the freedom to choose, but it is not actually students that make these choices. There is also some kind of education taking place, but this education neither takes the actual students into consideration nor

allows for transformative learning. Lynda Stone (2006) confirms this phenomenon when she investigates the effects of Enframing on research, asserting that
US graduate study in educational research entails a standardised over-emphasis
on methodology that ends up technologising research, reproducing knowledge
and limiting the possibilities for educational change (p. 541).

As it is then evident, educational critique ensuing from Heidegger's later philosophy of technology is obviously very critical for making us aware of technology's effects on our educational practices. It also highlights the general uniformity
that Enframing enforces, tending to swallow all potential differences. This critical
approach is also especially constructive, directing us to the poetic alternatives of
life, but this turn towards *poiēsis* coincides with an inherent *aporia*; indeed, what if
our world is already completely enframed? What if the impossibility of difference is
the way our world is constituted right now? What if we are too late? This is a matter in need of serious consideration. Heidegger (1977b) himself has stated, after
all, that '[t]he rule of Enframing threatens man with the possibility that it could
be denied to him to enter into a more original revealing and hence to experience
the call of a more primal truth' (p. 28). In consequence, this means that the possibility to be poetically engaged with the world might be a complete impossibility
for us. What, then, if this impossibility is the very reality that we are experiencing now? What if Enframing is the specific way that the world is structured and
defined? How can there even be space to think the alternatives? Furthermore,
how can education, which is concerned with the becoming and trans-forming of
the learned, offer new possibilities for connecting to the world? What if certain
technologised modes of relating to things are able to postpone the possibility of
all postponement? In other words, what if technology – being a totalising frame
of revealing – has already robbed the world from its potential for differentiation?

These questions do not concern only educationalists. They rather address the
essence of Heidegger's later critique of technology that, according to Feenberg
(2006), is based on sharp binaries and ahistorical descriptions that deprive technology of 'a socially and historically specific context and content' (p. 17). According to Feenberg, the discourse of Enframing appears to deny any possibility of
human agency, meaning, and by extension, the democratisation of technology.
As Feenberg (2006) puts it:

> Today it rages over the whole planet as a human deed: modern technol-
> ogy. But a universe ordered simply by the will has no root and no intrinsic
> meaning. In such a universe, man has no special ontological place but it is
> merely one force among others, one object of force among others. Meta-
> physics swallows up the metaphysician and so contradicts itself in the terrible
> catastrophe that is modernity. Heidegger calls for resignation and passivity
> *(Gelassenheit)*[2] rather than an active program of reform which would simply
> constitute a further extension of modern technology.
>
> (p. 184)

Feenberg (2006) raises here an important question that ultimately all theories of
radical constructivism or inventivism share; indeed, when the privileged human

being, which is traditionally endowed with the task of ordering the world and enforcing its will, is displaced from the driving seat, how are we supposed to imagine the possibility of moving forward? Are we to give up on the idea of driving entirely or is there a need for decision-making that urgently addresses the human being and that cannot simply be ignored? In order to start sketching what a Heideggerian response to this question might look like, we need to provisionally turn towards a supposedly later notion in Heidegger's thought, which is called *Gelassenheit* and is usually translated as *letting-things-be*. This notion, I argue, is not a delayed reflection in Heidegger's thought but one which is possible in later Heidegger precisely because early Heidegger made *Seinlassen* (*letting things be*) the basis of his phenomenological hermeneutics. This notion connotes a type of passivity, which is not to be defined in psychological terms and does not connote resignation as Feenberg suggests. *Seinlassen* is more than anything based on the realisation that the human being is a nearing being that both affects and is being affected by that which comes near. For this reason, I propose to investigate all the levels at which nearness is working, namely, the anthropocentric that offers, according to Ziarek (2000), a reconceptualisation of power relations, but also the non-metaphysical anti-subjectivist one describing a process of interaction between domains, which are not and cannot be fully known. In what follows, I go back to Heidegger's early discussions of metaphysics and technology in order to explain the way *nearness* becomes Heidegger's way of speaking about truth, learning, trans*formative* education, *paideia* and *Bildung*. Through this examination, it will be shown that educational matters are organically connected to Heidegger's own concerns about truth, technology and image.

3. Truth and education

According to Iain Thomson (2001), Heidegger attempts to do two things in his essay *Plato's Doctrine of Truth*: to 'trace the technologization of education back to an ontological ambiguity already inherent in Plato's founding ontological vision' and to 'show how forgotten aspects of the original Platonic notion of *paideia* remain capable of inspiring heretofore unthought possibilities of the *future* of education' (Thomson, 2001, p. 244). In order to do so, Heidegger (2009) delineates different notions of truth, even ones that remain 'unsaid' in the Platonic text, and addresses them in connection to the technologisation of Being (p. 155). For purposes of tracing this argument, we need to give first a brief account of the Allegory of the Cave, which, according to Socrates, the person recounting it, depicts the human condition as a movement from 'ἀπαιδευσία' – the lack of education – to 'παιδεία' – the state of being truly educated (Plato, n.d., 7.514a).

The story goes something like this: A group of people, prisoners from birth, are kept in a cave. Their legs and necks are shackled and the only thing they can see is the walls of the cave. Behind the prisoners, and on a higher level than their own, there is a fire and a small road. People walk in front of this fire, carrying things like statues that represent animals and people. Facing the shadows on the walls, the prisoners have come to know this imagery as the realm of the real (τὸ ἀληθὲς;

Plato, n.d., 7.515c). Finally, one prisoner is freed. They move around and even approach the fire that produced the fake images. According to Plato, this allows them to move closer to being and to see more correctly (μᾶλλον τι ἐγγύτερω τοῦ ὄντος καὶ πρὸς μᾶλλον ὄντα τετραμμένος ὀρθότερον βλέποι; Plato, n.d., 7.515d). The prisoner later ascends to the level of the cave's entrance and then moves into the open. There, after experiencing temporary blindness and seeing their reflections (εἴδωλα) in the water, they turn their gaze to the night sky. Finally, they look at the actual sun instead of the phantasms (φαντάσματα; Plato, n.d., 7.516b).

The analogies, usually detected between the story of the cave and Socrates' vision of being and education, are as follows: The world of the cave refers to the actual material things at which we look every day, whilst the world outside of the cave constitutes the realm of the eternal species (*eide*) and universal ideas (*idees*). These essences constitute the immaterial realm that allows us to perceive anything *as* something. For Socrates, the state of the uneducated person is similar to the state of the prisoner in the cave, in the respect that the person lacking education perceives the ephemeral copies of things – their shadows, idols and phantasms – without realising that they are not real. Following this analogy, education is the counterpart of the ascent into the luminous openness and the habituation to the new ways in which things manifest themselves. Education is a different way of seeing things; indeed, one that allows us to see them *as* they are, that is, in their ideal, universal and essential being. Furthermore, Heidegger (2009) asserts that true education entails the claim of responsibility and thus the return to the cave so as to liberate the other prisoners. In this way, Heidegger asserts that Plato depicts truth as a certain movement towards that which is 'more unhidden' – the 'ἀληθέστερον', whilst the true as ἀληθές 'fulfils the essence of παιδεία as a turning around' (p. 170). This reorientation 'can be achieved only in the region of, and on the basis of, the most unhidden, i.e., the ἀληθέστατον, i.e., the truest, i.e., truth in the proper sense' (ibid.).

As noted, Plato describes the first stages of this movement of truth as a kind of *nearness* (ἐγγύτης) that moves the one that sees towards the real being of beings, and this notion of nearness becomes the cornerstone of Heidegger's attempt to think Being. Nearness by no means presupposes the categorisation of beings and their representation, but rather the disposition to affect and to be affected by whatever comes near. The images Heidegger discusses either in connection to poetic language or to technological production or to the two distinct types of truth instantiate nearness as well.

Plato uses the word 'ὀρθότης' to describe the prisoner's look towards real things. He says that when the freed prisoner looks at the fire, he sees more correctly – ὀρθότερον. For correctness to exists, however, a certain measure is presupposed. Heidegger (2009) takes this measure to be what Plato calls *eide* or ideas, explaining that according to the Greek philosopher, '[t]he looks that show what things themselves are, the εἴδη, constitute the essence in whose light each individual being shows itself as this or that, and only in this self-showing does the appearing thing become unhidden and accessible' (pp. 169–170). However, what the Greek philosopher describes as εἶδος, namely, the 'visible form', and ἰδέα, as that which 'brings about presencing' suggests that '[a] being becomes present in each case in its whatness' (p. 173). 'Whatness' is what gets translated by the Romans as *essence*, namely, the unchangeable substance or the static core of a thing. It is, in other words, its presence, which is quite distinct from the movement into presencing that Heidegger (2000) attempts to sketch (p. 193). The belief in the 'whatness' of beings, and the eternal ideas that constitute their origin, transforms knowing into a form of representation that the subject needs to identify and categorise according to a predetermined idea. In this light, truth becomes the drawing of taxonomies, the finding of correspondences, between the universal and the particular, and the deciphering of likeness between a thing and its respective idea. Truth becomes *correctness* or what Plato calls the *orthon* (ὀρθόν).

This limited understanding of truth, Heidegger (2009) comments, inaugurates philosophy as metaphysics. Heidegger locates metaphysics' respective origin at a specific point in the Platonic text, explaining that: 'In the passage (516) that depicts the adaptation of the gaze to the ideas, Plato says (516 c3), 'Thinking goes μετ' ἐκεῖνα,"beyond" those things that are experienced in the form of mere shadows and images, and goes εἰς ταῦτα, "out toward" these things, namely the "ideas"' (p. 180). The word *μετά* denotes the movement beyond *physis* – beyond the actuality, materiality and particularity of natural beings, and into the realm of immaterial and eternal ideas, which is so central for *meta-physics*. Heidegger explains that according to this schema ideas constitute the

> suprasensuous, seen with a nonsensuous gaze; they are the being of beings, which cannot be grasped with our bodily organs. And the highest in the region of the suprasensuous is that idea which, as the idea of all ideas, remains the cause of the subsistence and the appearing of all beings. Because this "idea" is thereby the cause of everything, it is also "the idea" that is called the "good." This highest and first cause is named by Plato and correspondingly by Aristotle τό θεῖον, the divine. Ever since being got interpreted as ἰδέα, thinking about the being of beings has been metaphysical, and metaphysics has been theological.
>
> (pp. 180–181)

Metaphysics constitutes an ontotheological hierarchy that is based on the sharp distinction between beings and their essences, beings and Being, and indeed between difference and identity. For this reason, *eidos* 'that originally designates the visual aspect of a being now denominates what will, in fact, never be apprehended by physical eyes' (Sinclair, 2006, p. 29). This is not, however, the only distortion that Western metaphysics brings forth: indeed, according to Heidegger (1988), the Platonic theory of forms essentially trans*forms* the way the world is understood '*with a view to production*' (p. 106). In *The Basic Problems of Phenomenology* (1988), Heidegger explains that:

> What is formed is, as we can also say, a shaped product. The potter forms a vase out of clay. All forming of shaped products is effected by using an

image, in the sense of a model, as guide and standard. The thing is produced by looking to the anticipated look of what is to be produced by shaping, forming. It is this anticipated look of the thing, sighted beforehand, that the Greeks mean ontologically by eidos, idea. The shaped product, which is shaped in conformity with the model, is as such the exact likeness of the model.

(ibid.)

Conceptualised as such, production becomes the metaphysical structure that allows the understanding of everything and that elevates image at the rank of universal truth. On this point, Sinclair (2006) comments that 'Plato's separation of the *eidos* from the empirical being itself is motivated by the priority of the prototype or paradigm envisaged by the producer before the process of production' (p. 33). In other words, it creates what Michael Zimmerman (1990) calls 'productionist metaphysics', a framework that puts the human being at the centre, as the one receiving the image/idea and does the forming (re-presenting the image that the object needs to instantiate; p. xvi). In this light, production does become the paradigm of Being, whilst the human being is thought to be the being that gives Being.

In summary, Heidegger detects two types of truth in Plato: the first type, *alētheia*, refers to *the nearing of the unhidden*. It is what is revealed to the human being, that is, the being that slowly and painstakingly approaches truth through forms and images. The other type of truth, ὀρθότης, makes the human being the centre and measure of all things; it aspires to reveal degrees of correctness, likeness and difference. Translating this dichotomy in educational terms, we would have the following sketch: On the one hand, there would be the kind of knowing that corresponds to technological production; it presumes that the learner is the mediator of a type of knowledge that is in any case pre-given and predetermined. The learners' task is to comply with specific objectives and criteria and thus to fit the multiplicity of the world into normative ideas. This mediation would affirm that the world is known, familiar and always already understood. The use of rubrics that secure the objectivity of learning would be quite common with this perspective. On the other hand, truth as *alētheia* would suggest a kind of openness to the world. In this case, the learners are still the mediators of knowledge, but instead of categorising the world in accordance to pre-set standards, they let themselves be affected by that which is encountered. What's more, by opening themselves to the world, learners become aspects of this world, which are encountered by others. With such an approach, the students would be still able to learn, form and create, but they would also be responsive to that which is learned. Learning would be less structured and more guided by occasion; it would be both passive as to hear what is said and active as to attend to what is to be said.

This translation of types of truth into educational terms would be, however, premature at this point. It would, of course, be in accord with the standard Heideggerian approach through which educational matters are discussed, but it would also solidify a certain dichotomy between these two modes of knowing, namely, *alētheia* and *orthotis*, originary image and technological copy. In contrast, the purpose of this book is to deconstruct this binary by tracing the correspondences between its counterparts. In other words, the book will attempt to reveal things that remain unsaid in both the Platonic and the Heideggerian texts, namely, things concerning image and forms of knowing. In the next section, I attempt to flesh out this relation, which is central to nearness and suggestive of an alternative way of looking at Heidegger's philosophy of technology in connection to education.

4. Education and image

As already noted, Heidegger (2009) believes that the meaning of truth would reveal the true nature and proper aims of education. Education demands a reorientation towards real images and slow habituation to the new realities that these images allow to emerge. It is a process having 'to do with one's being and thus takes place in the very ground of one's essence' (p. 166). A mere choosing between possibilities, as indicated in the aforementioned discussion of further education, would be simply inadequate. A true turning around must take place as an existential and transformative dialogue, involving the individual and the world and allowing the individual to assume responsibility for this turning. In an attempt to explain the characteristics of this type of education, Heidegger turns to *Bildung*, a German word that traditionally means education and that bears the connotations of forming and formation. As Heidegger explains:

> On the one hand formation (*Bildung*) means forming someone in the sense of impressing on him a character that unfolds. But at the same time this "forming" of someone "forms" (or impresses a character on) someone by antecedently taking measure in terms of some paradigmatic image, which for that reason is called the proto-type [*Vor-bild*].
>
> (p. 166)

Heidegger's description of education as *Bildung* appears to bear no resemblance to his earlier exegesis of the Platonic definition of truth as *alētheia* and existential reorientation. In contrast, education sketched in this way seems to present profound similarities to the process of technological production, namely, a process that by no means considers the individuality of things, but rather traces correspondences between the things at hand and the pre-given standards set forth by ideal images. In other words, this specific description of education is closer to *orthotis* than *alētheia*. In order to investigate this ambivalence, we need to pay heed, however, to the way Heidegger (1988) describes the technological production of artefacts. Indeed, he says:

> The anticipated look, the proto-typical image (*Vor-bild*), shows the thing as what it is before the production and how it is supposed to look as a product.

The anticipated look has not yet been externalized as something formed, actual, but is the image of imagination (*das Bild der Ein-Bildung*), of fantasy, φαντασία, as the Greeks say – that which forming first brings freely to sight, that which is sighted. It is no accident that Kant, for whom the concepts of form and matter, morphe and hule, play a fundamental epistemological role, conjointly assigns to imagination a distinctive function in explaining the objectivity of knowledge. The eidos as the look, anticipated in imagination, of what is to be formed gives the thing with regard to what this thing already was and is before all actualization.

(p. 107)

This passage deserves scrutiny for many reasons: First, technological production appears to be understood through a Kantian lens and not through the metaphysical Platonic one. Second, the account of technological production foregrounds imagination as the one bringing together *morphe* and *hule*, universal and particular, sensuous and nonsensuous, and thus as that element opening up the possibility of escaping the demarcated oppositions of metaphysical thought. Third, it suggests that Heidegger himself reconceptualises *eidos* in order to free it from its metaphysical constraints and ultimately to sketch it as a type of imagination that connects, brings near and synthesises. Such reconsideration of the *eidos* (look, image) would suggest a process of formation that allows the individual and the world to come near. It would also underline the need for theoretical clarification of the *eidos* as a type of image. This will be addressed later in this book, through Heidegger's own reorientation and reinscription of Kantian imagination. For now, suffice it to say that Heidegger attempts to sketch a different kind of image, a poetic image. An image of this kind would be the Allegory of the Cave itself and also the image of the imprisoned human beings, which Socrates's interlocutor, Glaucon, describes as an 'ἄτοπον [. . .] εἰκόνα', an uncanny image, literally, an image without place (*topos*, Plato, n.d., 7.515a). Heidegger would say that the allegory is a poetic and, indeed, an originary image. In any case, Heidegger will eventually discuss poetic image in the text, ". . . Poetically Man Dwells . . .", and describe it as a process of welcoming the alien into the familiar and as a type of relatedness that lets something be seen, adding that 'poetic images are imaginings in a distinctive sense: not mere fancies and illusions, but imaginings [*Ein-Bildugen*] that are visible inclusions of the alien in the sight of the familiar' (Heidegger, 1975b, pp. 225–226).

The poetic/originary image is, according to Heidegger, the image that lets something be seen in contrast to copies and imitations. But, of what does this genuine image allow the sight? Is it the original presence (*Anwesenheit*) of things, their proper look or their respective essences? And does this mean that a representation of a thing, let us say, a painting of a pair of shoes, is some sort of a lesser entity, a degradation of the real thing precisely because it is nothing more than a copy? In *The Age of the World Picture*, Heidegger (1977a) seems to be headed towards this direction, equating representation with the essence of modern technology, with mathematical thinking, objectification, and calculation, asserting that

> [i]n distinction from Greek apprehending, modern representing, whose meaning the word *repraesentatio* first brings to its earliest expression, intends something quite different. Here to represent [*vor-stellen*] means to bring what is present-at-hand [*das Vor-handene*] before oneself as something standing over-and-against, to relate it to oneself, to the one representing it, and to force it, back into a relationship to oneself as the normative realm.
>
> (pp. 131–132)

Through the equation of the German *Vorstellung* with the Latin *repraesentatio*, Heidegger appears to juxtapose representation (*Vorstellung*) to originary presence (*Anwesenheit*), and, for this reason, Derrida (2007) attempts to deconstruct this opposition, arguing that:

> *Vorstellung* does not immediately seem to imply the meaning that is conveyed by the *re-* of *repraesentatio*. *Vorstellen* seems to mean simply, as Heidegger emphasizes, to pose, to dispose before oneself, a sort of theme on the theme. But this sense or value of being-before is already at work in "present." *Praesentatio* signifies the fact of presenting and *re-praesentatio* that of *rendering* present, of summoning as a power-of-bringing-back-to-presence.
>
> (p. 105)

Expanding on the argument, Derrida asserts that we cannot claim that there was once a prior Greek epoch during which things were purely present. Indeed, representation itself does not refer to the opposite of presence but to the very 'power-of-bringing-back, in a repetitive way' that is already central to the movement of *presencing* (p. 106). This means two things: that, on the one hand, *to present* means to 'bring to presence, into presence, cause or allow to come by presencing' and, on the other hand, 'because to cause or to allow to come implies the possibility of causing or allowing to return, then to render present, like all "rendering" and like all restitution, would be to repeat, to be able to repeat' (p. 106).

Heidegger, however, is not concerned with either the reference to real presences or with their singular instantiations. If the contrary was the case, Heidegger (1975a) would not have been able to see the true nature of Being in a Van Gogh painting depicting a pair of peasant shoes. Indeed, attempting to articulate the essence of the pair of shoes, Heidegger says that it is the painting, namely, a material re-presentation that becomes the site of truth, stressing that '[i]n the work of art the truth of an entity has set itself to work. "To set" means here: to bring to a stand' (p. 36). Certainly the painting is a form of representation and therefore a form of repetition, but its truth is neither derived from its sufficient likeness to a referent pair of shoes nor to its ability to repeat. In contrast, the painting itself constitutes the presencing of shoes, the site of truth, and the process of emergence.

Therefore, it seems that Heidegger does not aim to address the binaries, presence/ representation, production/reproduction, or even the identity/difference dichotomy. Rather, he discusses the process of presencing through the painting. To put this in different terms, Heidegger is preoccupied with nearness as presencing: the painting instantiates a modality that allows the world to come near in a certain way. This presencing does not force things to stand 'over-and-against' us as objects, usual familiarities, or consumable pieces of information. Rather, the painting unfolds as a poem, an image and a schema that includes the alien in the familiar. For this reason, I agree with Véronique M. Fóti (1985) who indicates that:

> Derrida takes for granted that a representation is a posited image or picture and, furthermore, that such an image may also be understood as a copy, substitute, scene, painting, or as the "objective reality" of an idea in Descartes's sense, so that metaphysical/epistemological, aesthetic, and political representation are held together *ab initio* under the aegis of the specular image.
>
> (p. 66)

Heidegger, in contrast, directs the discussion to a different direction, and even though his thought appears at times to devalue representation or at least some of its aspects, it would be rushed to think that Heidegger addresses representation in terms of repeatability, identity and difference. In contrast, what Heidegger (2002) attempts to do, is to think difference in terms of relation, belongingness and nearness. His emphasis on *presencing* points precisely to his constant concern with that which allows the togetherness of beings, through a play of presence and absence. With nearness, something comes near, only because something remains far. And, through this nearing of different things, something emerges instead of something else. Fóti (1985) therefore argues that Heidegger's critique

> cannot be understood on the model of a diminution or corruption of a primary presence (*Anwesenheit*) by representation, because *aletheia* is a presencing out of absence and into absence, and absence is not representable. Representative thinking does not presuppose, for its very possibility, a pregiven presence which it then breaks with and seeks to restitute; rather it first institutes the ideal of presence. It does so because it repudiates the togetherness of presence and absence, or what Heidegger calls the "sameness" (without identity) of showing and concealing.
>
> (p. 75)

Deciphering Being cannot be about the decoding of elements that make up a being, since, at their separate and autonomous states these elements cannot be the origin of these beings at all. Being is rather the synergetic nearing of elements and forces, allowing the presencing of beings. In this respect, the Heideggerian project is not so much concerned with difference, but rather looks to

> displace or inflect the economy of difference, and argue the need for a new mode of thinking relation: one that would be attuned to nearness rather than

difference, to the interval rather than the opposites, and to the transformative opening rather than negation.

(Ziarek, 2000, p. 134)

This displacement of difference, I would argue, does not seek to replace the
twofold movement of deferment and differentiation, but rather looks into considering the neglected synthesising and interpreting aspects that make any kind
of movement possible. In this respect, the question that Heidegger poses is this:
Which representations are capable of presencing, not in terms of difference or in
terms of likeness to an original presence, but in terms of nearness and emergence?
Can there be processes, which are formed by whatever is encountered and are
at the same time powers of formation? Can these structures receive measure and
constitute the measure? Can there be representations that let things be? Indeed,
can they make visible the alien in the familiar? And, in what ways, are these representations defined by language and technology? The preliminary distinctions,
I have sketched so far, refer to the representative and the poetic images, but we
do not really know a lot about these types of images. Because of this lack, Fóti
(1985) argues that

> [a]lthough Heidegger does indicate, in passing, that he considers the latter
> sense of *Bild* (poetic image) to be primary, he leaves the tension between the
> two senses quite unresolved. This is all the more surprising since both senses
> bring into play some of the linguistic complexities which Heidegger stresses
> with respect to the term "Bild." These comprise the links to the notions *of
> Bildung* as the education and formation of the person, of shaping and form-
> giving (*Bilden*), and of *phantasia* and imagining (*Einbildung*), which in turn
> is linked, through the cognate term *Vorstellung*, to representation, thetic
> positing, *Gestell*.
>
> (pp. 66–67)

The distinction between poetic and non-poetic image opens up a whole new discussion concerning types of images (*Bild*), types of forming (*Bilden*), their relation to imagination (*Ein-Bildungskraft*) – as the one being affected in receiving
and producing forms of imagining, and ultimately their connection with *Bildung*
as the very process concerned with human formation.

5. Conclusion

The important distinction between poetic/originary and non-poetic/representative/
derivative image will continue to unfold throughout this book while exploring
the aforementioned chain of notions related to image. Some of these notions
will be understood as being closer to language, while others will be related to
technological representation. Fóti (1985) made the point that Heidegger 'never
addresses the question of image directly' (p. 73). These *aporias*, however, sub-
sist throughout the Heideggerian work and underline the need for further discussion concerning the role of language and technology for the unfolding of

imagination, schematisation and nearness. This discussion will allow us to see if there is a possibility that a technologically and, even a mass-produced, image can be an originary image, and it will do so through Heidegger's understanding of image, imagination and nearness.

Nearness is understood here as a mode of knowing and connectedness to the world that allows the human being to participate in the unfolding of life through formative procedures. In this respect, nearness remains close to what Heidegger understood as imagination and time. Heidegger has, of course, hardly been theorised as a philosopher of imagination, but I believe that, as the critical and etymological excursion in *Plato's Doctrine of Truth* has shown, imagination becomes the forming process behind some of the most important Heideggerian notions while confirming that image, technology and language exist in a chain of concerns that need to be addressed together. In the next chapter, this connection is investigated by looking closely into Heidegger's phenomenological and hermeneutic methodology.

Notes

1 The term refers 'primarily to post-compulsory education in institutions which typically offer an extraordinary range of vocational and general education on a full-time and part-time basis. This extends to the education of adults. It does not normally include advanced academic work of the sort commonly associated with degree level (and beyond) in universities' (Standish, 1997, p. 458).
2 This is Feenberg's interpretation of the term, which will be discussed extensively in subsequent chapters in connection to phenomenology.

References

Derrida, J. (2007). The Retrait of Metaphor. In P. Kamuf & E. Rottenberg (Eds.), *Psyche: Inventions of the Other* (Vol. 1, pp. 48–80). Stanford, CA: Stanford University Press.

Feenberg, A. (2006). *Questioning Technology*. London: Routledge.

Fitzsimons, P. (2002). Enframing Education. In M. A. Peters (Ed.), *Heidegger, Education, and Modernity* (pp. 171–190). Oxford: Rowman & Littlefield Publishers, Inc.

Fóti, V. M. (1985). Representation and the Image: Between Heidegger, Derrida, and Plato. *Man and World, 18*, 65–78.

Heidegger, M. (1975a). The Origin of the Work of Art (A. Hofstadter, Trans.). In *Poetry, Language, Thought*. New York: Harper and Row.

Heidegger, M. (1975b). The Thing (A. Hofstadter, Trans.). In A. Hofstadter (Ed.), *Poetry, Language, Thought* (pp. 165–186). New York: Harper and Row.

Heidegger, M. (1977a). The Age of the World Picture (W. Lovitt, Trans.). In *The Question Concerning Technology, and Other Essays*. New York and London: Harper and Row.

Heidegger, M. (1977b). *The Question Concerning Technology and Other Essays* (W. Lovitt, Trans.). New York: Harper and Row.

Heidegger, M. (1988). *The Basic Problems of Phenomenology* (Revised ed.) (A. Hofstadter, Trans.). Bloomington: Indiana University Press.

Heidegger, M. (2000). *Introduction to Metaphysics* (G. Fried & R. Polt, Trans.). New Haven, CT: Yale University Press.

Heidegger, M. (2002). *Identity and Difference* (J. Stambaugh, Trans.). Chicago, IL: University of Chicago Press.

Heidegger, M. (2009). *Pathmarks*. In W. McNeill (Ed.). Cambridge: Cambridge University Press.

Peim, N., & Flint, K. J. (2009). Testing Times: Questions Concerning Assessment for School Improvement. *Educational Philosophy and Theory, 42*(3), 342–361.

Plato. (n.d.). Republic. *Perseus Digital Library*. Retrieved January 10, 2017, from www.perseus.tufts.edu/hopper/text?doc=Perseus:text:1999.01.0167

Sinclair, M. (2006). *Heidegger, Aristotle and the Work of Art: Poiesis in Being*. Basingstoke: Palgrave Macmillan.

Standish, P. (1997). Heidegger and the Technology of Further Education. *Journal of Philosophy of Education, 31*(3), 439–459.

Stone, L. (2006). From Technologization to Totalization in Education Research: US Graduate Training, Methodology, and Critique. *Journal of Philosophy of Education, 40*(4), 527–545.

Thomson, I. D. (2001). Heidegger on Ontological Education, or: How We Become What We Are. *Inquiry, 44* (3), 243–68.

Ziarek, K. (2000). Proximities: Irigaray and Heidegger on Difference. *Continental Philosophy Review, 33*(2), 133–158.

Zimmerman, E. M. (1990). *Heidegger's Confrontation with Modernity: Technology, Politics, and Art*. Bloomington, IN: Indiana University Press.

2 Hermeneutics and Heidegger's imaginings

1. Introduction

Phenomenology is a return to the things themselves. The hope to see things *as* the things that they are materialises in early Heidegger via a focus on *Seinlassen*, namely, the attitude that *lets things be*. This phenomenological approach that aspires not only to see and describe beings, but also to observe and describe the human being as a seer and a describer, is believed to be abandoned with Heidegger's later thought. Spiegelberg (1971), for example, comments that 'Heidegger himself has dropped all references to phenomenology from his later writings' and that phenomenology 'was fundamentally nothing but a phase in his development' (pp. 273, 349). This, however, is simply not the case: the *letting relationship* presents itself in different guises throughout Heidegger's thought, allowing him to imagine the nature of nearness. More specifically, Heidegger first describes phenomenology as a *letting-something-be-seen* while formulating his notion of authenticity as a *letting-something-be-heard*; finally, he transforms this *letting* disposition into a type of meditative thinking (*Gelassenheit* as *letting-things-be*; Heidegger, 1969). In what follows, I attempt to trace the different guises of *the letting relationship* that appears to be not simply a methodological approach but also the human relation to time. It is thus instantiated at times as the hermeneutic relationship and at other times as nearness and imagination. All these processes connote ways through which the human being relates to the world and ultimately experiences time.

2. Phenomenology in Heidegger

William Richardson (2003) studied most of Heidegger's oeuvre and wrote his now-classic book *Through Phenomenology to Thought*. Heidegger himself, responding to the author's request for a preface, suggested that the preposition *through* should substitute Richardson's original *from*, arguing that he understood phenomenology 'as the [process of] allowing the most proper concern of thought to show itself' (Richardson, 2003, p. xvi). This '*allowing*' or '*letting*', which bears anti-subjectivist, anti-productionist and non-metaphysical overtones, becomes the cornerstone of the thinking Heidegger is attempting to articulate throughout his prolific career, which began in 1907 with the study of Franz Brentano's 1862 dissertation titled *On the Manifold Senses of Being in Aristotle*, and which dealt explicitly with the question of the ontico-ontological difference (Sheehan, 1984). This attitude of *allowing* is also integral to what Heidegger will eventually come to understand as imagination and time, and we need, therefore, to go back to Brentano's thought so as to trace it.

Brentano's investigation led him to phenomenology and to the study of mental experience, professing in his *Psychology from an Empirical Standpoint* (1874) that 'all mental experiences is directedness toward or reference to a meant object [. . .] whether or not that object actually exists in the world', and concluding that: 'The essence of mental experience is intentionality – the minding-of-the-meant' (Sheehan, 1984, p. 291).[1] This means, according to Herbert Spiegelberg (1971), that 'Brentano for the first time uncovered a structure which was to become one of the basic patterns for all phenomenological analysis' (p. 41).

Brentano's further elaboration of intentionality led him to distinguish between three types of mental experiences believed to be instantiated as 'representations', 'judgements' and 'acts of love and hatred' (p. 42). Among the three types, Brentano thought that representation – with its 'relative simplicity, independence, and omnipresence in all psychological phenomena' was the most significant one (p. 43). He also claimed that representations are 'characterized by temporal modes' and that 'present time is given to us directly', whereas 'past and future times appear to us only by way of our present representations of ourselves as experiencing the past or as experiencing the future event' (p. 44). Even though Brentano made the important distinction between the immediacy of the present and the indirectness of the past and the future, he also talked about the 'original association' as the basic function of imagination '*Phantasie*'. According to Brentano, imagination brings the representations of the immediate past to the fore in an original and productive way so as to allow for the perception of a thing. The example given is of a 'pencil moving around in a circle' that I get to perceive not as a set of separated, isolated, or even simultaneous points in time, but as a whole continuous temporal movement. In this way, Brentano distinguished between imagination's 'acquired association', namely, the process in which imagination is involved when it brings back something from the past, and the aforementioned originary power that constitutes the thing *as* it is in the present (Mulligan, 2004, p. 78). As it then follows, phenomenology alludes, and indeed from its inaugural stage, and even despite its own intentions, to imagination's presence in perception.

Brentano's student, Edmund Husserl, diverged from his mentor's approach in an attempt to safeguard the centrality of the present. For Husserl, imagination works to represent the past, whereas perception allows the emergence of the present. In this respect, the moment, which has just passed is not yet part of the past; on the contrary, it is oriented towards the present and as such is involved in perception. In this way, Husserl – as I will explain at length in what follows – maintained the primacy of the *present*, which he attributed to the productive activity of perception, whereas he considered the past and the future to belong to the realm of reproductive imagination (Stiegler, 1998).

Husserl made many more contributions to phenomenology: in the *Logical Investigations*, he underlined that intentionality was a kind of 'directedness toward an object' instead of 'the object's immanence in consciousness', whilst he claimed that intentionality does not characterise all phenomena (Spiegelberg, 1971, p. 107). He then turned explicitly to

> the consciousness of the knowing subject to whom these phenomena appeared, i.e., in something he later came to call "transcendental subjectivity." Thus the "turn to the object" was supplemented by a "turn to the subject" by way of a new kind of reflection which left his erstwhile followers on the road to the "object" far behind.
>
> (p. 82)

Husserl (1984) then made the distinction between *simple* and *categorial intuition*, believing that simple intuition presents the object of perception 'immediately', whereas categorial intuition allows us to 'intend objects which cannot be intended in the simple founding acts, like "being red," "being a book"' (cited in Lahmar, 2006, p. 116). With the phrase, 'the paper is white', for example, we know to what the words 'paper' and 'white' refer, but we do not have direct knowledge of the content of the copula. In fact, we do not know what the *paper as being white* and the *paper as being paper* means. Sheehan (1984) explains:

> The dimension of the categorial corresponds to what the tradition called "being." More specifically, it is the "being-as" dimension of phenomena (X *as* Y = X *is* Y). Therefore, by freeing being from its mere status as a copula and by seeing it as a directly given phenomenon, Husserl showed that the full range of phenomenological immediacy covers not just entities but entities-in-their-mode-of-being, entities *as being* such or so. Intentionality in its full range is intrinsically ontological, a disclosure of entities in their being.
>
> (p. 291)

Husserl thus supported that we can 'obtain full adequate intuitive understanding' of beings that does not necessarily correspond to sense data. This meant that we can intuit 'general essences', which are the result of a certain process that relies on specific instances in order to reach a generalised 'ideation' (Spiegelberg, 1971, p. 118).

Following the example of his own teacher, Edmund Husserl, Heidegger objected the former's transcendental theory of consciousness and transformed intentionality into a kind of directedness grounded on the premise that a human being is not mere consciousness but an always already world-situated being oriented toward things that have meaning because of their own situatedness in the world. Heidegger, however, maintained the Husserlian belief that there *is* something that allows us to perceive things *as* things. For Heidegger, this *as-structure* is Being itself, and he proceeded to investigate it phenomenologically and kinetically, understanding it more like a movement that presences something out of absence, and less like a substance (Sheehan, 1984).

Leaving the Platonic 'whatness', essence or *idea* behind, Heidegger attempted to understand the way that the lack of sensuous origin allows the presencing of beings. In other words, Heidegger began to see Being as a temporal process, formulating in this way a version of phenomenology that affirms an openness that *lets beings be received*. What's more, Heidegger (2008) asserted that questioning, and especially questioning about Being, is one of the human being's modes of being. He thus claimed that 'the very asking of this question is an entity's mode of Being; and as such it gets its essential character from what is inquired about – namely, Being' (p. 27/28). Since for Heidegger, *Dasein*'s mode of being is temporal, and since Being is temporal as well, *Dasein*'s thinking is mostly conditioned by time. Heidegger explains this as follows: 'As understanding, *Dasein* projects its Being upon possibilities' (p. 188/148). Through this understanding, 'that which is understood – is already accessible in such a way that its "as which" can be made to stand out explicitly. The "as" makes up the structure of the explicitness of something that is understood. It constitutes the interpretation' (p. 189/149).

The 'as-structure' constitutes the openness that allows beings to be perceived, and it is 'grounded *in something we see in advance* – in a *fore-sight*' (p. 191/150). As such, it is also grounded in *Dasein*'s 'primordial state' as 'projection' (p. 192/151). This underlines the temporal constitution of *Dasein* and shifts the focus from the present onto the future, that is, onto something that is not here yet, an absence that allows something to come near. Heidegger explains:

> The way the Present is rooted in the future and in having been, is the existential-temporal condition for the possibility that what has been projected in circumspective understanding can be brought closer in a making-present, and in such a way that the Present can thus conform itself to what is encountered within the horizon of awaiting and retaining; this means that it must interpret itself in the schema of the as-structure [. . .]. *Like understanding and interpretation in general, the 'as' is grounded in the ecstatico-horizontal unity of temporality.*
>
> (p. 411/360)

At this moment, many interesting points come to the fore: First, the passage makes clear that, for Heidegger, the present is conjointly constituted by the future (awaiting) and the past (retaining), and this weaving of past memories with futural projections clearly highlights the role of absence for the manifestation of things. Second, the passage highlights the phenomenological exegesis of the procedural nature of Being. Third, it affirms that Being refers to the as-structure, a schema that is itself allowed by means of time *association*. Fourth, it equates a certain movement of bringing-close with the process of making present. And, finally, the fifth and perhaps the most important, point has to do with the striking resemblance between the Heideggerian temporality and the Kantian imagination. This connection is discussed in several Heideggerian texts, especially, in the *Kantbuch* and the *Basic Problems of Phenomenology*, whilst, in *Being and Time*, Heidegger (2008) declared that he intended to include 'Kant's doctrine of schematism and time, as a preliminary stage in the problematic of temporality'[2] in the first division of part 2 of *BT*[3] – which was, of course, never written (p. 64/40). For all these reasons, I discuss next the way Heidegger attempted to explain the surplus of meaning we receive in perception via a reinscription of imagination and temporality.

3. *Phantasia,* imagination, *Einbildungskraft*

Heidegger's (2008) *Dasein* does not possess an essence. Rather, 'it has its Being to be' (pp. 32–33/12). This being is *in* and *part of the world* as an evolving existence. Therefore, a phenomenological investigation of *Dasein* cannot be based on a fixed 'standpoint' or 'direction' but must be understood as a '*methodological conception*' that does not 'characterise the what of the object of philosophical research as subject-matter, but rather the *how* of the research' (p. 50/27). In this respect, Heidegger's methodology is an investigation that attempts to *let things present themselves as they are* instead of conforming them to what is determined beforehand. To put this in different terms, Heidegger's research reflects his own ontological vision concerning the human being and its modes of being.

Moving on to the explication of his methodology, Heidegger points out that phenomenology stems from the Greek words φαίνεσθαι (to show itself) and λόγος (discourse, language). Most importantly, he explains that λόγος (as discourse) has, according to Aristotle, the character of ἀποφαίνεσθαι; that is, it 'lets something be seen (φαίνεσθαι)' (p. 56/32). Since 'the λόγος is a letting-something-be-seen, it can be true or false', and, indeed be true not 'in the sense of "agreement"', but as ἀλήθεια – that is to say, in the originary way of revealing (p. 56/33). For Heidegger, this process cannot refer beforehand to any specific entity since part of the process is the discovery of the entity that needs to be investigated. He thus writes:

> 'Phenomenology' neither designates the object of its researches, nor characterizes the subject-matter comprised. The word merely informs us of the "*how*" with which what is to be treated in this science gets exhibited and handled. To have a science 'of' phenomena means to grasp its objects *in such a way* that everything about them which is up for discussion must be treated by exhibiting it directly and demonstrating it directly.
>
> (pp. 59/34–35)

If then the phenomenon is what shows itself from itself, and discourse (λόγος) is *the letting of what shows itself to show itself from itself*, then phenomeno-*logy*, to put it simply, aspires to describe beings as they emerge free from preconceptions. Description plays such a big role for the phenomenological process that Heidegger argues that the term 'descriptive phenomenology' is 'tautological' (p. 59/35). He then refers to the unbreakable bond existing between description and interpretation, explaining that:

> The λόγος of the phenomenology of *Dasein* has the character of a ἑρμηνεύειν, through which the authentic meaning of being, and also those authentic

structures of Being which *Dasein* itself possess, are *made known* to *Dasein's* understanding of Being. The phenomenology of *Dasein* is a *hermeneutic* in the primordial signification of this word, where it designates this business of interpreting.

(pp. 61–62/37)

As such, Heidegger's hermeneutics no longer designates the interpretation of a text but the interpretation of the world: *Dasein* is the interpreter of the world, and it lives by means of interpretation and language. In fact, language, for Heidegger (2008),

> [w]hen fully concrete, discoursing (letting something be seen) has the character of speaking [Sprechens] – vocal proclamation in words. The λόγος *is* φωνή, and indeed, φωνή μετά φαντασίας – an utterance in which something is sighted in each case.
>
> (p. 56/33)

The reference to Greek φαντασία, a term which is usually rendered in English as imagination bearing the connotations of fantasy and illusion, seems surprising here; it would seem that phenomenology as the method that allows beings to present themselves as they are, cannot be further from this notion, and therefore, John Sallis (1990) aptly comments:

> One ought not pass too easily over the paradox that is made to appear as soon as imagination is introduced into phenomenology. Set upon returning to the things themselves, phenomenology would appear to require just the opposite direction from that which imagination is believed typically to take. Phenomenology, it appears, could have nothing to do with those flights of phantasy, those fictions, with which imagination – seemingly oblivious to the things themselves – will have to do.
>
> (p. 97)

What are we to make then of this seeming contradiction? We need to investigate Heidegger's path into the exploration of imagination, which seems to draw here from alternative understandings of imagination, allowing him thus to transform *phantasia* into the cornerstone of *Dasein's* hermeneutical task and time. Indeed, the expression 'φωνή μετά φαντασίας' is one of Aristotle's five characteristics of reference (Sheehan, 1984). The word *phantasia*, of course, can be rendered as imagination but instead of connoting something like image, imitation, likeness or representation (all connotations of *phantasia's* Latin counterpart – *imago*), it connotes a process of bringing something into light similarly to *alētheia*, in that it shares the same φα-stem (light) with words like φαινόμενο (phenomenon) and απόφανσις (dictum). For this reason, Sallis (1990) maintains that in order to understand Heidegger's notion of imagination '[o]ne will need to come back to what Heidegger develops – even if largely in reference to the Greeks and to

Kant – as the basic problem of phenomenology. As a first move toward reinscribing imagination' (p. 97).

Martha Nussbaum's (1985) discussion concerning imagination's Greek interpretation is quite helpful here, since she argues, similarly to Heidegger, that one of the earliest meanings of the word *phantasia* comes from the verb *phainetai* (it appears). She also comments that in Plato's *Theaetetus* we find the following brief discussion: 'Is this "it appears" perceiving? . . . Then phantasia and perception are the same in the case of that which is warm and everything of that sort', which suggests that even sensuous data require imagination in order to be perceived as specific things (cited in Nussbaum, 1985, p. 242). Nussbaum then explains that Aristotle also maintains the connection between *phantasia* and perception, even though he uses the word in order to refer to representation. However, she asserts that 'the evidence indicates that his basic interest is in how things in the world appear to living creatures, what the creatures see their objects *as*' (p. 255). What's more, she claims that for Aristotle *phantasia* and *aesthesis* (feeling, emotion) are both present in perception, but whilst *aesthesis* is more passive in 'receiving perceptual stimuli', *phantasia* is active and perhaps transformative, allowing us to see something 'as a certain thing' (p. 259). This accordingly means that

> reception and interpretation are not separable, but thoroughly interdependent. There is no receptive "innocent eye" in perception. How something *phainetai* to me is obviously bound with my past, my prejudices, and my needs. But if it is only in virtue of *phantasia*, and not *aesthesis* alone, that I apprehend the object as an object, then it follows that there is no uninterpreted or "innocent" view of it, no distinction – at least of the level of form or object-perception – between the given, or received and the interpreted.
>
> (p. 261)

Understood from its Aristotelian origin, *phantasia* is the trans-*formative* element of perception that does not register reality as a neutral reproductive machine. In other words, elements like 'white' and 'paper' are not received separately from each other or as distinct properties of pre-existing beings, but rather are imaginatively synthesised unities of meaning. For this reason, Heidegger's task in *BT* is this: to reimagine imagination in terms of a knowing that is transformative and yet responsive to things. In order to do so, he has recourse to Kant's (2010) *Critique of Pure Reason*.

4. Heidegger's phenomenological interpretation of Kant

In Kant (2010), there are two faculties in cognition, endowed with different tasks. On the one hand, *sensibility* is responsible for receiving *sense data*, and on the other hand, *understanding* sorts out the manifold of data that is subsumed under the *categories*. Since the two faculties are quite different, Kant argues that 'there must be some third thing, which on the one side is homogeneous with the category, and with the phenomenon on the other, and so makes the application of the former to the latter possible' (p. 138). Continuing, Kant specifies that: 'This mediating representation must be pure (without any empirical content), and yet must on the one side be intellectual, on the other sensuous. Such a representation is the transcendental schema' (p. 139).

Mediating representations are not static images. They are *schemas* that belong to imagination. More specifically, productive imagination takes the manifold of intuitions, received in sensibility, and connects them with the use of schemas (procedural rules) that apply images to the pure *a priori* concepts (categories) of understanding, which are innate and independent from empirical reality. Kant calls this process *synthesis*, explaining that the term connotes 'the process of joining different representations to each other and of comprehending their diversity in one cognition' (p. 92). In contrast, *reproductive imagination* retrieves an already-formed temporal object, which has been synthesised by the productive transcendental imagination. In other words, the transcendental imagination 'is a condition of, not a result of experience', whereas 'reproductive' imagination is a recollection of experience (Furlong, 2002, pp. 116, 119).

The synthesis that takes place in transcendental imagination, called *reproduction*, is not the only synthesis taking place in cognition: Kant also mentions the synthesis of *apprehension*, which takes place in intuition (sensibility), and the synthesis of *recognition*, which takes place in the realm of the concept (understanding). However, imagination's significance lies in the fact that it is involved in the other two syntheses as well. Heidegger (1997) argues that *apprehension*, which refers to the reception of the *manifold of intuition*, is not possible without the mediation of a certain type of synthesis that reproduces and brings forth those intuitions that have just passed and subsequently connects them with those that are current. Without such synthesis, we would receive a muddle of disjoint intuitions and the 'mind would constantly fall from one phase of the now into another totally unconnected phase, in such a way that the earlier would simply be lost' (p. 238). In contrast, the mind is able of 'retainability and a repeated bringing forth of what is offered empirically' (p. 238). Through reproduction, imagination 'keeps open the *horizon of alreadyness*', retaining what is 'no-longer-now' in unity with the 'now' (p. 239). This 'pure synthesis of retaining constitutes the mind's being able to *distinguish* something like time' (ibid.). In other words, Heidegger equates the ability to retain and repeatedly access the *has-been*, which takes place in order to unify it with the 'now', with the possibility of time, arguing that '[t]o say "synthesis is related to time" is already a tautology' (p. 240).

Reproduction's role in apprehension underlines imagination's involvement in perception, and, according to Andrew Brook (2011), this is also Kant's position, since, in the second edition of his *Critique*, the philosopher proposes that apprehension is possible only via imagination. Heidegger, however, takes a step further, arguing against Kant that the third synthesis, namely, *recognition*, which takes place in the *concept*, is also related to time. Had Kant made such an assertion, he could have undone his sharp distinction between apprehension – implicated in intuition and thus in sensibility – and the synthesis of recognition – implicated in the concept[4] and thus in understanding. In order to proceed to such an undoing, Heidegger had to show that imagination is present not only in apprehension and in reproduction but also in recognition. Recognition is what brings all syntheses together, and Heidegger believed that by detecting the relatedness of all three syntheses to time and, in consequence, to imagination, he would have proven that Kant's insight, that there is a 'common root that lies 'beneath the two stems of knowledge (understanding and intuition)', was correct (Blattner, 2006, p. 167). Therefore, Heidegger (1997) admits that

> *in interpreting the third synthesis we go way beyond Kant*, because now the problem of the common root of both stems of knowledge becomes acute. We are concerned with understanding time and the I-think more radically and in the direction which is certainly visible in Kant, but which is not taken by him, i.e., in the direction of the synthesis of the power of imagination.
>
> (p. 243)

Recognition refers to the possibility of discerning and remembering earlier representations of a concept and associating these with current representations of the same concept (Kant, 2010). For Heidegger (1997), this synthesis refers to a process that enables both apprehension and reproduction precisely because it points to the fact that *what was retained and what was intuited belong to the concept*. For this reason, Heidegger argues that 'the designation "*recognition*" is quite misleading', proposing thus the term *identification*, which he believes would better explain the process of acknowledging an intuition as an instance of an *a priori* concept (p. 244). In this respect, identification would constitute a look into the past. This is not the case, however, since identification awaits for the unification of beings, and it thus 'opens up and projects in advance a whole – a whole which is in fact in one way or another disclosable and appropriatable in apprehension and reproduction' (p. 246). In this way, Heidegger (1997) transforms Kant's notion of *recognition* into the synthesis of '*advance awaiting* of a regional unity of offerable beings', or put differently, into the 'synthesis of *pre*-cognition', which constitutes an openness – given in advance – that lets beings show themselves (p. 256). This openness was to become the foundation of Heidegger's own understanding of temporality. Indeed, in *BT* the future has priority over the other dimensions of time; it is a kind of fore-sight that moves through the past, forming and projecting the present. Llewellyn (2000) comments:

> A pure synthesis of recognition is already involved in the subsequent return to a past now through the production of a pure reproduction of it. Such past futurity may be that of a projection I did not, but could have made. But I could not have made it without implying that I could now make a projection into the future future.
>
> (p. 42)

For Heidegger (1997), the constitution of the third synthesis indicates that all three syntheses are related to time, and this proves that both sensibility and understanding are related to time. Time allows for continuity, consistency and meaning (p. 247). To put it differently:

> The synthesis of *apprehension* is related to the *present*, the synthesis of repro-duction is related to the *past*, and the synthesis of *re-cognition* is related to the *future*. *Insofar as all three modes of synthesis are related to time and insofar as these moments of time make up the unity of time itself, the three syntheses maintain their unified ground in the unity of time.*
>
> (p. 246)

Heidegger then turns to *apperception*, namely, another process discussed by Kant, in order to elaborate further his account of recognition. For Kant (2010), apper-ception constitutes the possibility of all unity: it is that which allows the connec-tion between representations. Apperception refers to the 'I think' that allows me to 'join one representation to another' and to become 'conscious of the synthesis of them' (pp. 111–112). Kant adds:

> Consequently, only because I can connect a variety of given representations in one consciousness, is it possible that I can represent to myself the identity of consciousness in these representations; in other words, the analytical unity of apperception is possible only under the presupposition of a synthetical unity.
>
> (p. 112)

This conclusion allows Heidegger to move even further away from Kant. For Heidegger (1997), the apperception's ability to synthesise proves that 'the origi-nal synthesis of the three syntheses; that is, the unity of consciousness is in itself such a unity of pure time-related imaginative synthesis' (p. 277). This, in turn, is transformed into Heidegger's own structure of temporality. As he explains:

> But if the productive power of imagination is in this way nothing but the most original unity of the three modes of synthesis, then this power has essentially already unified in itself pure intuition and pure thinking, pure receptivity and pure spontaneity – or put more precisely, this power is the root which releases both from out of itself. The productive power of imagi-nation is the root of the faculties of subjectivity; it is the basic ecstatic con-stitution of the subject, of Dasein itself. Insofar as the power of imagination releases pure time from out of itself, as we have shown (and this means that the power of imagination contains pure time as a possibility), it is original temporality and therefore the radical faculty of ontological knowledge.
>
> (p. 283)

In this light, Heidegger manages to form a theory of temporality that refers to the interconnectedness of the three syntheses. These syntheses are themselves depended on an imagination that receives but also connects. Imagination thus lies at the heart of the unity of time, since temporality necessitates, above everything, connection and association. Sallis (1990) thus infers, that

> transcendental imagination can carry out the forming of time as the now-sequence precisely because it is identical with originary time, with temporality. Hence, in this move the identity of time and imagination is not only thought but now thought precisely *as identity*.
>
> (p. 108)

Since Heidegger's original aim is to reconstruct the notion of time, the identity of time and imagination makes imagination redundant. In fact, Heidegger (1973) argues that since 'transcendental imagination is transformed into something more originary', then "the designation 'imagination' becomes of itself inappropriate" (cited in Sallis, 1990, pp. 108–109). For this reason, it is important to remember that much of Heidegger's discussion on temporality is a discussion on imagination, namely, a type of imagination that constitutes the ground of 'receptivity' and 'spontaneity', and, above all, constitutes that which allows things to be.

5. Conclusion

At this early stage of Heidegger's thought, time and imagination intertwine and finally become identical: imagination forms (*Bilden*) time, and time is imagination's power to form (*Einbildungskraft*) (Llewellyn, 2000). In this respect, Heidegger (1997) understands imagination as

> *a forming and re-forming* [*Nachbilden*], that is, of making visible again the nows which have been; *of an in-advance forming* [*Vorbilden*], that is, of letting the now which is not yet present be sighted; and of forming an image [*Abbilden*], which brings the now which is present directly before or in front of us [*vorbilden*].
>
> (p. 282)

Taking this reorientation into account, image, formation, and imagination become indistinguishable from Heidegger's temporality. For Heidegger, time is the result of synthesis, an originary association that allows past, present and future to come together and give time. This original nearness of moments allows time consciousness and consciousness in general. Without this bringing-near of past and present, and presence and of absence, time cannot be formed or be identified as time by a subject that is able to observe its continuity and thus its own being in time.

The exegesis of imaginative formation and originary synthesis that creates time can very well refer to what Heidegger has sketched as originary/poetic image and which was precisely set in juxtaposition to technological representation. However, the defining feature of this opposition cannot be repeatability, as

Derrida would argue, since, according to Heidegger, repeatability as awaiting is a precondition for anything to exist in time and as time. Repeatability is just one aspect of the possibility to perceive time as time, identity and difference. Other aspects are nearness and association of moments. For Kant, this synthesis is based on *a priori* concepts that allow the categorisation of impressions. For Heidegger, this structure is transformed into the process of *in advance forming*, which is largely defined not only by that which has been but also by that which is to come through schematisation.

The forming power of this imaginative projection cannot be simply defined by what has immediately passed, or put differently: what has just passed relies itself on processes of formation that we need to explore. In this light, the important question rising here has to do with the nature of this *in advance forming*, and whether it is itself formed and affected by other processes and elements of formation. A preliminary answer would suggest that language plays a critical role for this process, since Heidegger equates the possibility of time with the potential of language. In *BT* Heidegger (2008) says:

> And only *because* the function of the λόγος as ἀπόφανσις lies in letting something be seen by pointing it out, can the λόγος have the structural role of σύνθεσις. Here "synthesis" does not mean a binding and linking together of representations, a manipulation of physical occurrences where the 'problem' arises of how these bindings, as something inside, agree with something physical outside. Here the συν has a purely apophantical signification and means letting something be seen in its *togetherness* [*Beisammen*] with something letting it be seen *as* something [*etwas als etwas sehen lassen*].
>
> (p. 56/33)

The emphasis on language's distinctive way of revealing underscores once more the role Heidegger ascribes to originary/poetic image, suggesting that language dominates this kind of synthesis. However, there is also a possibility that language's deictic powers constitute the mere surface of imagination's deeper forming power, and, for this reason, this power needs to be investigated further. The chapter that follows does precisely that.

Notes

1 Physical phenomena, in contrast, lack such directedness (Spiegelberg, 1971).
2 The implications of this connection will be discussed in the next section.
3 From here on, *Being and Time* will be referred to as *BT*.
4 These concepts or categories refer to quantity, quality, relation and modality.

References

Blattner, W. (2006). Laying the Ground for Metaphysics. In C. B. Guignon (Ed.), *The Cambridge Companion to Heidegger* (pp. 149–176). Cambridge: Cambridge University Press.

Brook, A. (2011). Kant's View of the Mind and Consciousness of Self. *The Stanford Encyclopedia of Philosophy.* Retrieved August 12, 2012, from http://plato.stanford.edu/archives/win2011/entries/kant-mind/

Furlong, J. E. (2002). *Imagination.* London: Routledge.

Heidegger, M. (1969). *Discourse on Thinking* (J. Anderson & H. Freud, Trans.). New York, London, Toronto and Sydney: Harper & Row.

Heidegger, M. (1997). *Phenomenological Interpretation of Kant's Critique of Pure Reason* (P. Emad & K. Maly, Trans.). Bloomington: Indiana University Press.

Heidegger, M. (2008). *Being and Time* (J. Macquarrie & E. Robinson, Trans.). Oxford: Blackwell.

Kant, I. (2010). *The Critique of Pure Reason* (J. M. D. Meiklejohn, Trans.). Happy Valley, PA: The Pennsylvania State University.

Lahmar, D. (2006). Categorial Intuition. In H. L. Dreyfus & M. A. Wrathall (Eds.), *A Companion to Phenomenology and Existentialism* (pp. 115–126). Oxford: Blackwell Publishing Ltd.

Llewellyn, J. (2000). *The HypoCritical Imagination: Bettwen Kant and Levinas.* London and New York: Routledge.

Mulligan, K. (2004). Brentano on the Mind. In D. Jacquette (Ed.), *The Cambridge Companion to Brentano* (pp. 66–97). Cambridge: Cambridge University Press.

Nussbaum, M. (1985). *Aristotle's De Motu Animalium.* Princeton, NJ: Princeton University Press.

Richardson, W. J. (2003). *Heidegger: Through Phenomenology to Thought.* New York: Fordham University Press.

Sallis, J. (1990). *Echoes: After Heidegger.* Bloomington, IN: Indiana University Press.

Sheehan, T. (1984). Heidegger's Philosophy of Mind. In G. Floistad (Ed.), *Contemporary Philosophy: A New Survey* (Vol. 4, Philosophy of Mind, pp. 287–318). The Hague: Nijhoff.

Spiegelberg, H. (1971). *The Phenomenological Movement: A Historical Introduction.* The Hague: Nijhoff.

Stiegler, B. (1998). *Technics and Time, 1: The Fault of Epimetheus* (R. Beardsworth & G. Collins, Trans.). Stanford, CA: Stanford University Press.

3 Imaginative synthesis as metaphor

1. Introduction

What does it mean for a being to be in time; indeed, be a being by means of the association that gives both time and the understanding of time? What's more, what does it mean for a being to be defined by the *in advance forming* of time, a temporal projection into the future itself formed by the past and the present's unfolding? The specific temporal constitution of *Dasein* has many different implications, but most importantly, it suggests that *Dasein* is conditioned by death. Death is, according to Heidegger (2008), the point where all the *in advance forming* leads, and where it essentially ends, explaining that:

> *Only an entity which, in its Being, is essentially* **futural** *so that it is free for its death and can let itself be thrown back upon its factical "there" by shattering itself against – that is to say, only an entity which as futural, is equiprimordially in the process of* **having-been,** *can, by handing down to itself the possibility it has inherited, take over its own thrownness and be* **in the moment of vision** *for 'its time'. Only authentic temporality which is at the same time finite, makes possible something like fate – that is to say, authentic historicality.*
>
> (p. 437/385)

This condensed passage encompasses much of Heidegger's early philosophy of time. It shows that Heidegger's reinscription of Kantian imagination unfolds both at the individual and the collective realms – both of them connected through *Dasein*'s possibility to inherit a past, which is not its own but rather its community's, and through *Dasein's* opportunity to reappropriate this past so as to form its future. *Dasein's* indebtedness to its community's past is the reason that Heidegger explains that *Dasein*'s 'own past – and this always means the past of its "generation"– is not something which *follows along after Dasein*, but something which already goes ahead of it' (p. 41/20). This possibility is determined by a distinct existential reference to *Dasein*'s opportunity to live an authentic life, and this possibility will prove to be an essential point of differentiation between modes of being that either promote authentic relations or inauthentic ones. The most decisive moment for the possibility of living an authentic life is the moment of vision, and we turn now to a careful examination of it.

2. The moment of vision as poetic image

Heidegger asserts that *Dasein* is the site of truth, in the respect that it allows truth to shine by offering space for this shining. If this space is not cleared, and if it is not open, then *Dasein's Da* – literally, its *there* – will be potentially formed by ways of being, which are chosen by others and not by *Dasein* itself. In other words, a *Dasein* living inauthentically allows itself to be formed by concerns, which are not really its own and not chosen by it. These concerns come rather from the domain of public sphere, and thus Heidegger maintains that:

> In utilizing public means of transport and in making use of information services such as the newspaper, every Other is like the next. This Being-with-one-another dissolves one's own Dasein completely into the kind of Being of 'the Others', in such a way, indeed, that the Others, as distinguishable and explicit, vanish more and more. In this inconspicuousness and unascertain-ability, the real dictatorship of the "they" is unfolded. We take pleasure and enjoy ourselves as *they [man]* take pleasure; we read, see, and judge about literature and art as *they* see and judge; likewise we shrink back from the 'great mass' as *they* shrink back; we find 'shocking' what they find shocking. The "they", which is nothing definite, and which all are, though not as the sum, prescribes the kind of Being of everydayness.
>
> (p. 164/127)

Through this passage, but also consistently throughout his work, Heidegger juxtaposes the possibility of authentic time to the reality of mundane time conditioned by technologies of transportation, communication and information. The possibility, however, of becoming an authentic self is part of *Dasein's* temporal constitution and thus cannot be erased. Heidegger explains that 'the Being of *Dasein* means ahead-of-itself-Being-already-in-(the world) as Being-alongside (entities encountered within-the-world)' (p. 237/193). This corresponds to the three ecstasies of time that constitute *Dasein's* being: *Dasein* derives itself from the *future* while being constituted from some *past* situation or other along with other beings and involvements in the *present*. The three ecstasies are formed into unity by '*care*' (*Sorge*) shown either towards things (concern) or towards another *Dasein* (solicitude) (p. 238/194). It is thus through concern and solicitude that *Dasein* moves towards a project that is in the future, whilst being itself affected by the has-been in the experience of the right-now.

This futural projection that, similarly to Kantian recognition, allows past and future to be involved in the present, has a twofold nature: it is constituted by the determinateness of death and the indeterminateness of life. *Dasein* can choose at any moment to be one thing or another. The decision to take on a single possibility is realised through 'anticipatory resoluteness', which is the state of '*Being towards* one's ownmost, distinctive potentiality-for-being' (p. 372/325).

Heidegger explains that this possibility is based on the structure of the *letting relationship*, which is inherently futural:

> This letting-itself-*come-towards*-itself in that distinctive possibility which it puts up with, is the primordial phenomenon of the *future as coming towards*. If either authentic or inauthentic *Being-towards-death* belongs to *Dasein's* Being, then such Being-towards-death is possible only as something *futural* [. . .].
>
> (p. 373/325)

Guilt is another important element of authenticity: *Dasein* needs to feel guilty of allowing its life to drift aimlessly away so as to finally experience anxiety in the face of its impending death. In this respect, the finality of death becomes at any time the possibility of experiencing life authentically. The realisation of death, as the unavoidable end, allows the consideration of new beginnings: *Dasein* realises that death is the possibility of eliminating all other possibilities and thus finally experiences the 'burden' of having to choose for itself. In order to do so, *Dasein* needs to suspend some possibilities and choose some others and thus to differentiate itself from its current non-self – that is to say, from its undifferentiated 'they-self'. As evident from the above phrasing, a temptation to read *Dasein* in terms of différance is quite strong here, in fact, so much so, that the philosopher Bernard Stiegler (1998) theorises *Dasein's* temporal structure in precisely these terms, asserting that:

> *Dasein* is the being who differs and defers [*l'étant qui diffère*]. A being who differs and defers should be understood in a twofold sense: the one who always puts off until later, who is essentially pro-jected in deferral, and the one who, for the same reason, finds itself originarily different, indeterminate, improbable. The being who defers by putting off till later anticipates: to anticipate always means to defer. *Dasein* has to be: it is not simply – it *is only* what *it will be;* it *is* time. Anticipation means being-for-the-end. *Dasein* knows its end. Yet it will *never* have knowledge of it. Its end is that toward which it is, in relation to which it is; yet its end is what will never be *for Dasein. Dasein is for* the end, but its end *is not* for it. Although it knows its end absolutely, it will always be that in relation to which it will never know anything: the knowledge of the end always withdraws, is concealed in being deferred.
>
> (p. 231)

Stiegler's reading reveals here an identification of *Dasein's* constitution with *Derridean* différance. Before addressing this point, however, we need to focus on Heidegger's own discussion of authenticity, which involves a value judgment with important implications: *Dasein* can at any time choose to be different, but it cannot simply choose to be authentic. Sheer will might procure difference, but it will not procure authenticity. For authenticity to occur, *Dasein* needs to hear the call of consciousness. In fact, Heidegger (2008) states that *Dasein* has the opportunity to be its own true self through the 'call of conscience' to which 'there corresponds a possible hearing', explaining that '[o]ur understanding of that appeal unveils itself as our wanting to have a conscience [*Gewissenhabenwollen*]' and thus underlying that an element of desire and acceptance comes into the constitution of the call (p. 314/270). It is perhaps, for this reason, that Llewellyn (2000) argues that the structure of the call takes us back 'to the root of Kant's *Achtung*, attentive respect or reverence', which also refers to imagination's possibility to receive transformatively (p. 38). Indeed, it seems that imagination plays a decisive role for receiving the call by allowing the synthesis and interpretation of the call *as* call. Therefore, Heidegger (2008) says that '[t]he call comes *from* me and yet *from beyond me*' (p. 319/275). This is possible since I am never just me and only me. I am always already a process of unfolding. I am always already here and there: here in the present as an heir of the past and here in the present as a projection into the future. This process is never complete and never linear. At any moment, I am becoming this being, which I am, and that being, which is different from what I have been. I am never entirely present to myself. Heidegger expands on this:

> In its "who," the caller is definable in a "worldly" way by *nothing* at all. The caller is *Dasein* in its uncanniness: primordial, thrown being-in-the-world as the "not-at-home" – the bare "that-it-is" in the "nothing" of the world. The caller is unfamiliar to the everyday one-self; it is something like an *alien* voice. What could be more alien to the "one," lost in the manifold "world" of its concern, than the Self which has been individualized down to itself in uncanniness and been thrown into the "nothing"?
>
> (pp. 321–322/276–277)

The rift, caused by the alien approaching the familiar and the uncanny coming close to the homely, takes us back to the poetic image, straight in the depths of Plato's cave, and right in the neighborhood of the 'ἄτοπον [. . .] εἰκόνα', in other words, close to the image without place that constitutes precisely the realm of allegory and metaphor. What does this mean, however? What do the similarities, shared by the call of consciousness, the poetic image, and the allegorical Platonic tale, tell us about the moment of vision? First, they propose that similarly to true education, which we discussed in the first chapter, authenticity relies on a *nearing process* that allows the familiar and the alien to approach each other, not in terms of negotiating difference, but in terms of weaving associations that may possibly lead to the emergence of newness, the play of presence and absence, the ascent into the light, and *alētheia*. A complete break with familiarity is neither possible nor desirable; such a break would create an absolute rapture between the self that I currently am and the self towards whom I move. This rupture would connote the complete loss of a self. In contrast, an uncanny image, an *atopos eikona*, allows the distortion of the familiar, the disruption of the everyday, and the breakdown of automated responses so as to create new ways of seeing. This process instanti-ates the constitution of seeing something *as* something, which is a prerequisite for authenticity, namely, for the possibility to see a future non-self as myself, to see what is currently absent and yet imaginable. It is thus a hermeneutical process that imagines something, which is not there: it brings this not-there into existence and offers it a place. As such, it cannot be put in propositional terms. In contrast, *nearing*, just like the originary image, is less of a designation and more of a meta-phor, an irony, and a paradox. In this respect, it appears that *Dasein* is a being of différance, as Stiegler (1998) interprets it to be, but its existence cannot be thought only in terms of deferment and of differentiation but also in terms of the imaginative nearness that allows the constitution of differentiated beings through synthesis. Différance necessitates synthesis and *hermeneia*. Maybe, it is precisely because of this possibility – emergence through nearness – that Heidegger under-lines language's possibility to reveal, arguing that

> [t]he 'Being – true' of the λόγος as ἀληθεύειν means that in λέγειν as ἀποφαίνεσθαι the entities *of which* one is talking must be taken out of their hiddenness; one must let them be seen as something unhidden (ἀληθές); that is, they must be *discovered*.
>
> (pp. 33/56–57)

In what comes next, we turn to Heidegger's own poetic alien and uncanny lan-guage in order to investigate in greater depth the nature of the poetic originary allegoric image and its connections to language. The questions we aim to answer are the following: What types of linguistic processes contribute to *Dasein's in advance forming*? What possible constitutive elements of language – if any – allow it to become a movement of nearness? And, finally: is the possibility of nearing confined to language, or is it open to other instantiations?

3. Heidegger's language and metaphor

Heideggerian phenomenology does not constitute a methodology that a human being employs in order to express and possibly to impose meaning on things. Phenomenology is rather meant to be the revealing of the revealing; it is the let-ting be seen of the letting be seen. That the language of Heidegger's particular phenomenology is to appear uncanny and unhomely, enveloped in poetic and mystical phrasing, is not to be thought as an accident, but rather as the recourse that a human interpretive being has so as to escape metaphysics. This much desired deviation is bound then to be a deviation from the philosophical lan-guage that constitutes itself the metaphysical conceptual and Platonic context of Western thinking.

However, Heidegger's precise escape from metaphysics is deeply metaphorical and undeniably metaphorical by means of spatial phrasing. The notion of *nearness*, which is discussed here as one of his central concerns, is paradigmatic of this lan-guage. But, of course, it is not the only one; for Heidegger, the human being is best defined as a *being-there*, a *being-in-the-world*, a *site* for itself and Being, a *projection into* the future and *an in advance forming*. There is then something to be said about Heidegger's language and spatial metaphorics; indeed, something to be said about a philosophy of time that speaks almost exclusively through spatial terms, allowing Paul Ricœur (2004) to proclaim that '[t]he constant use Heidegger makes of metaphor is finally more important than what he says in passing against metaphor' (p. 331).

Before delving deeper into the rich, metaphorical and almost foundational structures of Heidegger's philosophy, we need to address first what Heidegger has to say explicitly about metaphor. Of course, someone could soon argue that philosophy itself is not possible without metaphor; indeed, no concept is possible without metaphor, or as Derrida (2007) puts it: 'there is nothing that does not go on without metaphor and through metaphor' (p. 50). Why would then the philosophical subject named, Martin Heidegger, be any different in his performance and articulation of thought? Why would the investigation of the Heideggerian text prove to be anything more than a case study, representative of all philosophy and of all thinking? The answers to these questions have to do with both the Heideggerian intention and the Heideggerian text, namely, a text attempting to rethink thinking and to do so metaphorically and in opposition to metaphor. In this respect, the Heideggerian text creates the metaphoric *aporia* par excellence – *aporia* being already a spatial metaphor – and clears the path for that which can be said about metaphor and for that which can be said in general. Truly, as Derrida (2007) beautifully puts it, what is at stake with the Heideggerian text is 'the apparently metaphoric power of a text whose author no longer wishes that what happens in that text and what claims to get along without metaphor there be understood precisely as "metaphoric"' (p. 64).

So, now, as we have promised, let us turn to a conspicuous excerpt from Lecture 6 of *The Principle of Reason* (1955/56) where Heidegger (1996) discusses the interpretive nature of *aesthesis* since, for him, sensing something can never be a mere registering of sense data. In other words, Heidegger believes that we never sense only with the help of our sense organs, since a certain interpretation has to take place in order to sense something *as* something. This accordingly suggests that in sensing there is no firm separation between sensuous and nonsensuous. Sensing is always already a form of interpretation and metaphorical perceiving. Heidegger, however, argues that sensing cannot be thought as metaphorical, since metaphor constitutes precisely the essence of the distinction between sensuous and nonsensuous. In his words:

> Because our hearing and seeing is never a mere sensible registering, it is also off the mark to insist that thinking as listening and bringing-into-view are only meant as a transposition of meaning, namely, as transposing the supposedly sensible into the nonsensible.
>
> The idea of "transposing" and of metaphor is based upon the distinguishing, if not complete separation, of the sensible and the nonsensible as two realms that subsist on their own. The setting up of this partition

of the sensible and nonsensible, between the physical and nonphysical is a basic feature of what is called metaphysics and which normatively determines Western thinking.

(p. 48)

*Meta*physics, literally the movement *meta* (beyond) *physis* (nature), seeks to establish a separation between that which is perceived by the senses (*aistheton*) and that which is thought (*noeton*). The belief in this separation and concomitant transposition is probably hidden in some way behind Aristotle's take on metaphor as the 'the application of a strange term either transferred from the genus and applied to the species or from the species and applied to the genus, or from one species to another or else by analogy' (Aristotle, n.d., 1457b7–9). Metaphor is thought as the improper transposing of a name, which properly designates a sensuous object, over to a nonsensuous being. This transposition allows, according to Heidegger (1996), the solidification of a sharp distinction between the realms of sensible and non-sensible and thus ultimately the construction of the metaphysical structure of Western thinking. Heidegger's belief that metaphor and metaphysics refer to the same thing will continue to inform his work up until the end. His comment in *The Nature of Language*, for example, that Honderlin's line, '[n]ow for it words like flower leaping alive/must find', that articulates the likeness between words and flowers, should not be treated as metaphor – as this 'would mean that we stay bogged down in metaphysics' – is indicative of this persisting conviction (Heidegger, 1982, p. 100). However, different philosophers, like Jacques Derrida (Derrida & Moore, 1974), Paul Ricœur (1978) and Giuseppe Stellardi (2000), agree on the fact that Heidegger's own understanding of metaphor is way too limited and conventional precisely because the separation between the sensuous and the non-sensuous need not be the defining characteristic of metaphor. More specifically, Stellardi (2000) proclaims that

> Heidegger accepts without problem the traditional notion of metaphor, and consequently condemns it with no possibility of appeal. Metaphor, whether living or dead, is always for him caught up in the circle of representation, technique [*Technik*], and metaphysics.
>
> (p. 140)

That metaphor should be found in the same web of signification with metaphysics and technology is repeatedly asserted by Heidegger, believing himself that Platonic metaphysics foregrounds truth as agreement and, more precisely, as agreement between a technologically produced representation and its respective *eidos*. Metaphor is thought to be entangled with metaphysics precisely as the very process that solidifies the distinction between the particular copy (sensuous) and the eternal idea (nonsensuous), and grounds technological production – that is to say, the process that connects the realm of objects with the realm of ideas – as the paradigm of Being. This identification will eventually become the keystone of Heidegger's effort to destruct metaphor and metaphysics. However, the condemnation of metaphor should not essentially be the path towards the destruction of metaphysics; an alternative route could have been opened up through the reconceptualisation of *aesthesis* as inherently metaphorical. Heidegger, as already indicated, takes the first step on this path, but his restricted understanding of metaphor does not allow him to go any further. Indeed, discussing the role of the senses for thinking, Heidegger (1996) explains that

> [w]hatever is heard by us never exhausts itself in what our ears, which from a certain point of view can be seen as separated sense organs, can pick up. More precisely, if we hear, something is not simply added to what the ear picks up; rather, what the ear perceives and how it perceives will already be attuned and determined by what we hear [. . .]. Of course our hearing organs are in a certain regard necessary, but they are never the sufficient condition of our hearing, for that hearing which accords and affords us whatever there really is to hear. The same holds for our eyes and our vision. If human vision remained confined to what is piped in as sensations through the eye to the retina, then, for instance, the Greeks would never have been able to see Apollo in a statue of a young man or, to put this in a better way, they would never have been able to see the statue in and through Apollo.
>
> (pp. 47–48)

In other words, the manifestation of the world, which served at a certain point of time as the only way the world presented itself to the ancient Greeks, was not a constructed fiction or an illusion placed over objective reality. Rather, the world appeared (φαίνεσθαι) in this and only in this way. The world as phenomenon, Heidegger seems to argue, is perceived with the assistance of both *aesthesis* (the senses) and *phantasia* (imagination), or better yet: *aesthesis* perceives imaginatively and through the modification of sense data. This realisation would, of course, suggest not only that Heidegger is in agreement with the Aristotelian theory of perception but also with the Kantian description of imagination as an associative power. But, is this the case? And, what implications would such an alliance have on Heidegger's philosophy of *hermeneia*?

The imaginative and transformative nature of perception serves well Heidegger's effort to overturn the metaphysical distinction between the sensuous and the nonsensuous. This exposition, however, would have also constituted a well-suited argument concerning perception's reliance on metaphor conceptualised as pertaining to language and cognition. In support of this intuition, we can focus on a unique moment of hesitation, in *Being and Time (BT)*, during which Heidegger (2008) remarks that 'λόγος is just *not* the kind of thing that can be considered as the primary ' "locus' of truth", adding that, in contrast, *sense* or

> [α]ἴσθησις, the sheer sensory perception of something, is 'true' in the Greek sense, and indeed, more primordially than the λόγος [. . .]. Just as seeing aims at colours any αἴσθησις aims at its ἴδια (those entities which are genuinely accessible only *through* it and *for* it); and to that extent this perception

is always true. This means that seeing always discovers colours, and hearing always discovers sounds. Pure νοεῖν is the perception of the simplest determinative ways of Being which entities as such may possess, and it perceives them just by looking at them. This νοεῖν is what is 'true' in the purest and most primordial sense; that is to say, it merely discovers, and it does so in such a way that it can never cover up.

(pp. 57–33)

Noein, namely, the function of *nous* – the minding of the mind – refers to cognition. It is a synthesis allowing things to appear as things, and it is closely connected to the as-structure and Being itself. As such, it could be the case that *noein* is not defined – or, at least, not fully defined – by language. On this point, we can, in fact, insert the pertinent comments made by Giambattista Vico (1668– 1744), an Italian philosopher who believed that human beings first communicated 'by means of signs and gestures', and that '[m]etaphor was then the primary mode of knowing and understanding the world' (Modell, 2003, p. 15). According to Vico, metaphor allowed the animistic and mythological understanding of the world and constituted a cognitive capacity that preceded language (ibid.). Vico's insight is quite impressive, but his belief that '[e]very metaphor is a fable in brief' could also allow us to make a strong connection with the Heideggerian text (1744, cited in Modell, 2003, p. 15). The point at which Heidegger explains the Greek worshiping of Apollo is quite crucial in this respect. After all, the thinking that allowed the Greeks to perceive Apollo while looking at the statue, and indeed perceive the statue as Apollo himself, would not have been possible without the mediation of the statue and the process of its production. In consequence, this suggests that perception is mediated through exteriors figures, just like the statue, but it is also mediated via processes of figuration that join the sensuous with the non-sensuous, even before any distinction between the two is made.

It is precisely this point that is of so much interest here: indeed, the degree to which, if anything at all, metaphorical and imaginative processes of perception are conditioned by exterior images. By using the words *image* and *figure*, I refer here to both figures of speech and artefactual figures that affect imagination and perception. Actually, if this point could be proven, then metaphor would be accordingly reconfigured as a process of schematisation that both language and technology instantiate. In order to examine this possibility, we need to look into the transcendental imagination once again, presenting precisely the way *aesthesis* functions as metaphorical synthesis and in accord with the Kantian imagination and the Heideggerian temporality.

A quick objection to this argument would be articulated in the following terms: Kantian imagination forms unities by weaving together the sensuous and the nonsensuous. The latter realm pre-exists in the mind and in the form of categories, whereas *aesthesis*, understood as metaphorical synthesis, denotes a process that is not determined *a priori* but is constituted through its own conditions of possibility; indeed, generating itself through its response to exterior images that ultimately co-constitute its rules of association, pretty much the same way that the statue conditioned the ancient Greeks' perception of their world. It could also be argued that even if this is what Heidegger's intuition expressed, namely, that imagination as synthesis is conditioned by exterior images, this is not in any case in accord with the Kantian text. Lastly, it could be stated that even if we prove that this is the nature and the role of imagination in Heidegger, its associative process cannot be ascribed – not even partially – to metaphor. In search of answers, I turn next to Bernard Stiegler's work on schematism.

4. Schema, metaphor and exteriorisation

As already noted, Kant (2010) argues that the productive imagination works through the schema – a third thing that connects the manifold of appearances in intuition with the discrete categories in understanding. Schemas need to be understood as both receptive and active: they require being able to receive anything, to connect everything and to be nothing in particular. Schemas have to receive the exterior by actively turning it into the interior, and they are innate; they are not produced by experience. In contrast, they are transcendental or pure, that is, empty of all empirical content. As such, 'the schema is clearly distinguishable from the image', and, for this reason, the schema has priority over the exterior image (p. 141). Kant (2010) explains:

> The schema is, in itself, always a mere product of the imagination. But, as the synthesis of imagination has for its aim no single intuition, but merely unity in the determination of sensibility, the schema is clearly distinguishable from the image. Thus, if I place five points one after another [. . .] this is an image of the number five. On the other hand, if I only think a number in general, which may be either five or a hundred, this thought is rather the representation of a method of representing in an image a sum (e.g., a thousand) in conformity with a conception, than the image itself, an image which I should find some little difficulty in reviewing, and comparing with the conception.
>
> (p. 141)

Commenting on this particular Kantian position, Stiegler (2011) argues that a concept like a number presupposes a certain kind of *exteriorisation* that makes the very concept of number possible. Consciousness, he maintains, needs to be '*transferred* onto external supports, as prosthetic memory as well as fetishes of the imagination and "projection screens" for all its phantasms' (pp. 53–54, emphasis added). Stiegler calls these supports tertiary retentions, which refer to Husserl's notion of exteriorised memory in the form of images. This, however, needs further explanation.

For Husserl, a temporal object like a melody is constituted around the current note (primal impression) that is preserved as *primary retention* and that is combined with that which is to come, namely, *protention*. These categories resemble the Heideggerian ecstasies of time and the Kantian syntheses constituting consciousness. However, these theories diverge with regards to the significance they assign to each dimension. For Husserl the *now* is the absolute site of experience and "the source-point of all temporal positions" (Husserl, 1991, cited in Brough & Blattner, 2006, p. 128). Both retention and protention are oriented towards the now and retention 'is the actual holding on to what has elapsed as it moves away from the now', whereas protention is understood as 'a moment of the actual phase of the ongoing perception that immediately opens me up to further experience, usually of what I am presently experiencing, without running through it in advance as if it were present' (Husserl, 1991, cited in Brough & Blattner, 2006, pp. 128–129). In this respect, both retention and protention are attached to the now and constitute perception without any mediation on the part of representation. Such exclusion of representation leads to imagination's exclusion from perception, which is precisely what Husserl intended, since the opposite case would have rendered any distinction between perception and imagination impossible (Stiegler, 1998). For Husserl, perception and imagination are not only distinctly different from each other but also from 'tertiary' memory – called 'image-consciousness' – referring to externalised representations/images in the form of works of art and any other type of simulacra. He writes:

> Perception . . . is the act that places something before our eyes as the thing itself, the act that *originarily constitutes* the object. Its opposite is *representation* [*Vegegenwartigung, Representation*], understood as the act that does not place an object itself before our eyes but precisely represents it; that places it before our eyes in image, as it were, although not exactly in the manner of a genuine image-consciousness . . . If we call perception the *act in which all 'origin'* lies, the act that *constitutes originarily*, then *primary memory* is *perception* . . .
>
> (Husserl, 1991, cited in Stiegler, 1998, p. 248)

Derrida (1973), however, comments that despite Husserl's desire to constitute an absolute pure present and thus an absolute presence, absence – relating to retention as time elapsed – and also relating to protention as time to come – is already part of the present time, presence and perception. He writes:

> One then sees quickly that the presence of the perceived present can appear as such only inasmuch as it is *continuously compounded* with a nonpresence and nonperception, with primary memory and expectation (retention and protention). These nonperceptions are neither added to, nor do they oc-*casionally* accompany, the actually perceived now; they are essentially and indispensably involved in its possibility. Husserl admittedly says that retention is still a perception. But this is the absolutely unique case – Husserl never recognized any other – of a perceiving in which the perceived is not a present but a past existing as a modification of the present [. . .].
>
> (p. 64)

Absence is always already participating in presencing, and Stiegler (2011) argues that imagination unfolds through this play of presence and absence, involving exterior supports that contain memories. Stiegler explains, for example, that when we listen to a temporal object for the second time – a repetition made possible precisely because this music is retained in exterior technological support – the memory of the first hearing becomes our secondary memory and modifies the impression we get when we listen to the melody for the second time. In this way, reproduction – and, in consequence, imagination, as that which brings forth passed impressions and constitutes representations – participates in perception. No two hearings (or repetitions) can be the same, since during the second hearing 'I hear from the position of an expectation formed from everything that has already musically happened to me – I am responding to the Muses guarding the default-of-origin of my desire, within me' (Stiegler, 2011, p. 19). This reliance on past memory, however, can be extended to the first hearing, since we can imagine that even the first hearing of a melody can be interpreted through whatever, musically or otherwise, has happened to us. Therefore, what is at stake is not simply the repetition but the original synthesis that makes perception possible as a kind of nearing of notes, moments and memories, constituting something as a temporal object. Stiegler (2011) offers in what follows another pertinent example, through Kant's discussion of the concept of the number, commenting that:

> A number always in some way presupposes a capacity for tertiary retention – whether via children's fingers, a magician's body, an abacus, or an alphanumeric system of writing – which alone can facilitate numerization and objectification. This capacity has a history, during which at one point the *concept* of one thousand (1000) became possible. Properly understood, this conception is first and foremost a process. Until a certain point quite recent relative to the long history of humanity, the number 1000 was literally inconceivable to a human consciousness without the *tools* for thinking it, when 1000 ("one thousand," or the figure/image . . . or IIIIIOI000) had not yet been *elaborated*.
>
> (p. 51)

In this respect, the schema is not only a process but also a product of a process that is not to be found in the individual, but rather in the intermediary process of synthesis that joins together the individual to its historical time and to exterior space. This necessity of spatialisation, figuration, and transfer does not evade Kant (2010), who points out that:

> We cannot cogitate a geometrical line without drawing it in thought, nor a circle without describing it, nor represent the three dimensions of space without drawing three lines from the same point perpendicular to one another. We cannot even cogitate time, unless, in drawing a straight line (which is to serve as the external figurative representation of time), we fix our attention on the act of the synthesis of the manifold, whereby we determine successively the internal sense, and thus attend also to the succession of this determination.
>
> (pp. 124–125)

As Rudolf A. Makkreel (1990) notes, Kant, in the *Second Deduction*, describes transcendental productive imagination as 'figurative' without sufficiently explaining the change of terminology. However, Makkreel asserts that 'the term "figurative" aptly suggests *the graphic, more spatial qualities* that the imagination contributes to synthesis. Insofar as the imagination synthesises it serves the understanding, but in the role it also brings to bear some of its own formative power' (p. 30, emphasis added). Spatialisation, though, often understood as specific to language – spatial metaphors attest, after all, to time's metaphorical transformation as space in language – can also be reconceptualised as a general process of metaphorical exteriorisation in matter, be it in sound – as it happens with language – or in any other kind of matter. In what comes next, some conclusions are drawn.

5. Conclusion

Paul Ricœur (1978) suggests that we think metaphor in terms of predication instead of denomination. In the utterance, for example, 'Juliet is the sun', what is at stake is not the transfer of a word, namely, the noun *sun* to an improper position, but rather a certain way of seeing Juliet *as* the sun, or seeing Juliet in connection to the sun, or even seeing Juliet in analogy to the sun. In Ricœur's (1978) reconceptualisation of metaphor, the focus is transferred from the word, or even the sentence, to discourse itself. In this view, metaphor's main function is not substitution but nearness, that is, a kind of nearness that brings together distinct domains, allowing them to be associated to each other. For this reason, Ricœur points out that metaphor is close to Kantian schematism, since both schematism and metaphor are 'a thinking and a seeing'[1] (p. 147). Especially, 'metaphor by analogy', Ricœur comments, is a way of *seeing as*, and this, we can add, immediately evokes Heidegger's Husserlian influences and the origin of his quest to reinscribe imagination, temporality and Being through the *as-structure* (p. 148). It is no wonder then that Ricœur (2004) comments that above everything, 'the "place" of metaphor, its most intimate and ultimate abode, is neither the name nor sentence nor even discourse but the copula of the verb to *be*. The metaphorical "is" signifies both "is not" and "is like"' (p. 6). For Ricœur (1978), the essence of metaphor lies in its ability to bring two completely different terms close to each other, and through predication, whilst he argues that 'predicative assimilation' allows the two terms to become 'semantically proximate' in a movement of 'rapprochement' that confirms 'a typical kinship [between metaphor and] . . . Kant's *schematism*' (1978, p. 148). Expanding on this, Ricœur (1978) explains that metaphor 'is nothing else than this move or shift in the logical distance, from the *far* to the *near*' (p. 147: emphasis added).

These same metaphors of space and, above all, nearness that allow Ricœur to rethink metaphor, become Heidegger's means of describing what it is to be-*in-the-world* and what it is to be an interpretive being in the world, namely, a being that receives other beings not as muddles of isolated data but *as* certain beings. *Nearness*, *remoteness* and *distancelessness* allow Heidegger to draw connections between time, truth, language, thinking, and technology. These metaphorical terms should then be understood less as the fanciful play of language and more as the free play of imagination unfolding in language and in other media. Indeed, if metaphor lies at the heart of Heidegger's notion of nearness, temporality and imagination, can we still infer that the power of schematisation is restricted to linguistic play? Our investigation up to this point has shown that schemas are mediating powers affected themselves by materiality and exteriority, and we can thus assume that exterior figures of matter and exterior figures of speech affect and instantiate imagination. In pursue of definite answers, the next three chapters investigate nearness as presented in Heidegger's thought.

Note

1 See also Hausman (1989).

References

Aristotle. (n.d.). Poetics. Retrieved January 10, 2017, from www.perseus.tufts.edu/hopper/text?doc=Perseus%3Atext%3A1999.01.0056%3Asection%3D1447a

Brough, J. B., & Blattner, W. (2006). Temporality. In H. L. Dreyfus & M. A. Wrathall (Eds.), *A Companion to Phenomenology and Existentialism* (pp. 127–134). Oxford: Blackwell Publishing Ltd.

Derrida, J. (1973). *Speech and Phenomena* (D. B. Allison, Trans.). Evanston: Northwestern University Press.

Derrida, J. (2007). The Retrait of Metaphor. In P. Kamuf & E. Rottenberg (Eds.), *Psyche: Inventions of the Other* (Vol. 1, pp. 48–80). Stanford, CA: Stanford University Press.

Derrida, J., & Moore, F. (1974). White Mythology: Metaphor in the Text of Philosophy. *New Literary History, 6*(1), 5–74.

Hausman, C. R. (1989). *Metaphor and Art: Interactionism and Reference in the Verbal and Nonverbal Arts.* Cambridge, New York, Melbroune: Cambridge University Press.

Heidegger, M. (1982). The nature of language. In *On the Way to Language* (P. D. Hertz, Trans.). Oxford: Harper One.

Heidegger, M. (1996). Lecture Six (R. Lilly, Trans.). In *The Principle of Reason* (pp. 1–49). Bloomington: Indiana University Press.

Heidegger, M. (2008). *Being and Time* (J. Macquarrie & E. Robinson, Trans.). Oxford: Blackwell.

Kant, I. (2010). *The Critique of Pure Reason* (J. M. D. Meiklejohn, Trans.). Happy Valley, PA: The Pennsylvania State University.

Llewellyn, J. (2000). *The Hypo Critical Imagination: Bettwen Kant and Levinas.* London and New York: Routledge.

Makkreel, R. A. (1990). *Imagination and Interpretation in Kant: The Hermeneutical Import of the Critique of Judgment.* Chicago, IL: University of Chicago Press.

Modell, A. H. (2003). *Imagination and the Meaningful Brain.* Cambridge, MA: MIT Press.

Ricœur, P. (1978). The Metaphorical Process as Cognition, Imagination, and Feeling. *Critical Inquiry, 5*(1), 143–159.

Ricœur, P. (2004). *The Rule of Metaphor: The Creation of Meaning in Language* (R. Czerny, K. McLauughlin & J. Costello, Trans.). London and New York: Roudtledge Classics.

Stellardi, G. (2000). *Heidegger and Derrida on Philosophy and Metaphor: Imperfect Thought.* New York: Humanity Books.

Stiegler, B. (1998). *Technics and Time, 1: The Fault of Epimetheus* (R. Beardsworth & G. Collins, Trans.). Stanford, CA: Stanford University Press.

Stiegler, B. (2011). *Technics and Time, 3: Cinematic Time and the Question of Malaise* (S. Barker, Trans.). Stanford, CA: Stanford University Press.

4 The ready-to-hand

Nearness in early Heidegger

1. Introduction

Heidegger talks about the human being as a thoroughly situated existence defined by the *in advance forming of itself*. This synthesis of projection, effected by the orientation towards the future and by the looking back at the past – necessitates an investigation into the 'when' and the 'where' of its taking place. In other words, we need to look into the place in which *Dasein* as *being-there* becomes *a there*. Such an investigation, of the taking place of *Dasein*, would have led us towards an exploration of that which we call space. The Heideggerian project, however, is more interested in that which makes the world appear *as* world, and it thus looks into new ways of understanding space. According to Heidegger (2008): 'A glance at previous ontology shows that if one fails to see Being-in-the-world as a state of *Dasein*, the phenomenon of worldhood likewise gets *passed over*' (p. 93/65). By paying heed to the '*unitary* phenomenon' of 'Being-in-the-world', Heidegger advances the belief that the world is possible, only because *Dasein* is in the world. The emphasis here is on the word *unitary* since for Heidegger the phenomenon of *worldhood* cannot 'be broken up into contents which may be pieced together', but still it can be studied in its 'constitutive items', which refer to other structures such as the 'in-the-world', the 'worldhood' of the world, *Dasein* as the '*entity* which in every case has Being-in-the-world as the way it is' and, last but not least, '*Being-in*' (pp. 78–79/53). In this light, the question that Heidegger needs to answer is this: What is it like for a being like Dasein to be *in* something like the world? Heidegger, therefore, explains that 'being-in' does not mean 'a spatial "in-one-another-ness"' and that the word *in* does not

> primordially signifies a spatial relationship of this kind. "In" is derived from *"innan"* – "to reside", *"habitare"*, "to dwell" [sich auf halten]. '*An*' signifies "I am accustomed", "I am familiar with", "I look after something." It has the signification of "*colo*" in the senses of "*habito*" and "*diligo*".
>
> (2008, p. 80/54)

Heidegger (2008) then detects in the very etymology of '"*ich bin*" ["I am"]' the meaning of residing and dwelling, namely, the very structures that he considers synonymous with being and knowing (ibid.). He then argues that '[k]nowing the world (νοεῖν) – or rather addressing oneself to the "world" and discussing it (λόγος) – thus functions as the primary mode of Being-in-the-world, even though Being-in-the-world does not as such get conceived' (p. 85/59).

Heidegger's remarks here pave the way for our understanding of his notion of *being-in*, which shifts any supposedly self-evident focus from the concept of mathematical space, and reiterates existence in terms of understanding, knowing, dwelling and caring.

In general, the structure of *being-in* is very important for philosophy and it has been equated early on with existence. Aristotle, for example, makes clear that all things – except for some special entities, like the heavens, gods, numbers and points – exist in a place somewhere. In contrast, things that lack a place or are not found in a *topos* lack existence (Casey, 1997). As Aristotle (2012) states in the fourth book of *Physics,* 'things which exist are somewhere (the nonexistent is nowhere – where is the goat-stag or the sphinx?' (p. 1). Heidegger, however, is precisely interested in the things that lack place and thus exist as 'ἄτοπος [. . .] εἰκόνα', namely poetic images that speak with the power of absence and that cause disturbances in what we thought to be familiar. Therefore, it comes as no surprise that Heidegger's existential analytic, which is articulated with the help of terms like *being-in, dwelling, homeliness* and *unhomeliness, nearness* and *distance,* plays with the binaries of absence and presence, and familiarity and unfamiliarity that Heidegger sees as indispensable for poetic revealing.

What is of the utmost importance for the argument formulated in this book, however, is the fact that *Dasein* is a *being-in-the-world* in terms of using tools or *letting these tools be involved* in and as *Dasein*'s time. However, this time, which is mediated by instruments, is understood in opposition to authentic time, which is constituted through the determinateness of the imminent human death, and, of course, through language's *apophantic* properties. Creating such an opposition, leads to Heidegger's restricted understanding of *Dasein*'s historicality since both the *in advance* part and the *forming* part of *Dasein's in advance forming* are limitedly understood through the care shown for that which is being formed instead of through the care for the very processes conditioning this forming. From this perspective, aspects of existence like materiality, embodiment, spatiality and prostheticity are not understood as participating in *Dasein's forming* process, probably because these dimensions of being are associated with the technological exteriorisation of metaphorical processes in perception. In what follows, I discuss the steps Heidegger takes in order to eliminate the technological aspects of being from his theorisation of authentic time.

2. The ready-to-hand and the construction of the already-there

In attempting to articulate *Dasein's being-in-the world*, Heidegger understands space, contrary to Kant, not as an attribute of the mind in the form of a category, but as that which results from 'our active being and our practical involvements in the world' (Arisaka, 1995, p. 4). Without a doubt, space is, for Heidegger, lived through the 'thinghood' of the objects *Dasein* uses and through the interconnections *Dasein* experiences with these things and with other *Dasein*s. Theodore Kisiel (1993) explains:

> The field of objects which yields the original sense of being is that of the *pro-duced* object accessible in the course of usage. Accordingly, it is not the field of things in their theoretical reification but rather the world encountered in going about our producing, making, and using which is the basis, the accord-ing-to-which and toward-which of the original experience of being [. . .].
>
> (p. 264)

Dasein encounters things like ancient Greeks encountered 'πράγματα – that is to say, that which one has to do with in one's concernful dealings (πρᾶξις)'. For Heidegger, 'the specifically 'pragmatic' character of the πράγματα' is *equip-ment* that 'is essentially 'something in order to' ["'etwas um-zu . . .''']' (2008, p. 97/68). Taking this explanation into consideration, anything – a pen or a car or a spaceship – is something used in order for one of my projects to take place: a pen is used in order to take notes concerning the groceries I intend to buy, a car is what I use in order to get to work and a spaceship is what I build in order to find myself in some other point in space. In each one of these cases, the tool appears to be not a mere tool, a physical some*thing* or some material being, but rather a medium that conveys my concern towards my projects and that transfigures my time as a certain type of time. This concern is what participates in the forming of my future self and indeed delivers me to it.

The 'in-order-to' of equipment is, according to Heidegger, translated into dif-ferent variations of purposiveness, such as 'serviceability, conduciveness, usability, manipulability' and '*assignment* or *reference* of something to something' (ibid.). This purposive and referential nature of equipment allows tools to constitute expandable networks: my pen refers to my notebook, in which I make my grocery list, and the list points to the car, which is going to take me to the supermarket, and the car refers to the road system, and so on. In this respect, tools and their networks constitute our *already-there* that conditions our existence by presenting the world with certain possibilities and certain restrictions. I cannot, for example, choose a road that is not built, and I might actually never be able to imagine such a possibility. Heidegger calls this structure 'readiness-to-hand' whilst believ-ing that *Dasein* interacts with the ready-to-hand by *letting it be involved* in the unfolding of its temporality by means of a special kind of sight, namely, circum-spection (p. 98/69). As he describes it:

> If we look at Things just 'theoretically', we can get along without under-standing readiness-to-hand. But when we deal with them by using them and manipulating them, this activity is not a blind one; it has its own kind of sight, by which our manipulation is guided and from which it acquires its specific Thingly character. Dealings with equipment subordinate themselves to the manifold assignments of the 'in-order-to'. And the sight with which they thus accommodate themselves is *circumspection*.
>
> (p. 98/69)

Circumspection is the average pretheoretical way in which we perceive the world. It is an integral function of perception, permeating our dealings with technology. As a distinct sight located in our 'practical behaviour', it is quite different from other modes of relatedness to things (p. 99/69). Indeed, Heidegger explains how theoretical sight works in different ways:

> This kind of Being towards the world is one which lets us encounter entities within-the-world purely in the *way they look* (εἶδος), just that; *on the basis* of this kind of Being, and *as* a mode of it, looking explicitly at what we encounter is possible. Looking *at* something in this way is sometimes a definite way of taking up a direction towards something – of setting our sights towards what is present-at-hand. It takes over a 'view-point' in advance from the entity which it encounters. Such looking-at enters the mode of dwelling autonomously alongside entities within-the-world. In this kind of '*dwelling*' as a holding-oneself-back from any manipulation or utilization, the *perception* of the present-at-hand is consummated.
>
> (p. 89/62)

The theorisation of the world according to image, or as Heidegger puts it as present-at-hand, refers to the way we think about things when we do not have them at hand, but when we look at them from a distance. It is a scientific attitude that objectifies things or even a type of awareness that presents itself when we realise that a tool is 'missing', is broken or unavailable. In such a case, the tool enters the 'mode of *obtrusiveness*' and 'reveals itself as present-at-hand and no more' (p. 103/173). Circumspection, conversely, is concerned with the manipulation of things, whilst permitting their withdrawal into inconspicuousness. This is because during our everyday dealings 'that with which we concern ourselves primarily is the work – that which is to be produced at the time' and not the actual tools we handle (p. 99/70).

However, the fact that Heidegger asserts that it is the care for that which is to be done that constitutes *Dasein's* time, and not the substratum that allows such concern to take place, has important consequences for *Dasein's being-in*, which is theorised as a state of mere temporal involvement. Indeed, Heidegger explains that '[i]f something has an involvement, this implies letting it be involved in something' (p. 115/184). This involvement is grounded in the discovery of the purpose for which things as ready-to-hand have been freed and not in their own particular constitution. In other words, through the use of the keyboard I am using right now, I have already discovered the intention of writing a text and of becoming a writer, and it is specifically this concern that defines and constitutes my "*wherein*", which '*is the phenomenon of the world*'. In consequence, 'the structure of that to which [*woraufin*] *Dasein* assigns itself is what makes up the *worldhood* of the world', and this is what *Dasein* experiences as nearness (p. 119/186). In fact, Heidegger, explains that the ready-to-hand refers to entities,

> which are 'close by' [in der Nähe]. What is ready-to-hand in our everyday dealings has the character of *closeness*. To be exact, this closeness of equipment

has already been intimated in the term 'readiness to hand', which expresses the Being of equipment. 'Every entity that is "to hand" has a different closeness, which is not to be ascertained by measuring distances. This closeness regulates itself in terms of circumspectively "calculative" manipulating and using'.

(p. 135/102)

This identification of technology with closeness and calculative thinking comes to form the core of Heidegger's critique of modern technology. Indeed, modern technology is criticized, in his later writings, for denying us the possibility to experience nearness. In order to explicate this further, we need to understand what Heidegger (1969) calls calculative thinking. For Heidegger this type of thinking

consists in the fact that whenever we plan, research, and organize, we always reckon with conditions that are given. We take them into account with the calculated intention of their serving specific purposes. Thus we can count on definite results. This calculation is the mark of all thinking that plans and investigates. Such thinking remains calculation even if it neither works with numbers nor uses an adding machine or computer. Calculative thinking computes. It computes ever new, ever more promising and at the same time more economical possibilities. Calculative thinking races from one prospect to the next. Calculative thinking never stops, never collects itself.

(p. 46)

A stark contrast then arises between poetic/originary/nearing image and technological/calculative/representative image, indeed, one replacing the dichotomy between authentic and inauthentic time, *alētheia* and *orthotis*. However, in order to fully understand these new binaries, we need to go deeper into Heidegger's critique of technology and into his philosophy of nearness, which becomes all the more focused on time.

3. The public, the prosthetic and the historical

As explained in the previous section, technology possesses a referential nature: tools do not exist isolated either from each other or from their users. The tool I use refers me to another tool that is involved in a project with which some other *Dasein* is also engaged. In this respect, both of us share the same inherently relational and social space. In fact, Dreyfus (1991) comments that this is an aspect of equipment's public character, since

[e]quipment displays *generality* and obeys *norms*. First, a piece of equipment is the equipment it is no matter who uses it. Hammers, typewriters, and buses are not just for me to use but for others too. Equipment is for 'Anybody'– a general user [. . .]. Second, there is a normal (appropriate) way to use any piece of equipment.

(p. 51)

Equipment's fundamentally social character 'directs attention to the way in which being-there is as much a being-with others (*Mitsein, MitDasein*) as it is a being amidst or being-alongside things (*Sein bei*)' and illustrates how societies are 'established and constituted through the organization of space' (Malpas, 2007, pp. 87–88). Specific communities have specific spatialisations of social interactions, configured by technologically produced means. In light of this, Heidegger (2008) discusses the 'work-world of the craftsman', asserting that in the work-activity the craftsman encounters, even if indirectly, 'those Others for whom the "work" ["Werk"] is destined', that is, the wearers for whom the shoes are made (p. 153/117). Malpas (2007) points out that this means that

> involvement with others is organized and oriented through this equipmental structure, which is also a social structure. And to the extent that the ordering of the world of equipment is something laid out 'in space,' so too the ordering of the social is a spatial ordering.
>
> (p. 88)

One would suppose that the necessary fusion of the technological, spatial and social realms that Heidegger's analysis seems to presuppose, would also underline the basic role of the socio-technological substrata for the conditioning of *Dasein*'s unfolding and its *in advance forming*. Heidegger, however, has characterised this space as the exclusive arena of the undifferentiated *they-self*. Indeed, if a tool needs to be for everybody, then everybody is turned into the *nobody*, experiencing a type of nearness to the world that Heidegger considers to be inauthentic. Such nearness constitutes a constant movement between ends, but these ends are not important when set in comparison to the one and absolute end of death. Death is the end that makes authentic life possible, and according to Stiegler (1998), this true '"finality" consequently appears to precede the possibility of an already-there of the what-ready-to-hand' (p. 249). Stiegler strongly objects this Heideggerian claim, arguing that *Dasein* cannot simply be a pure self-affected entity that relates to the world authentically, through the possibility of its own death, since the 'finality (being-toward-the-end) and the facticity of an *already-there* are inextricable' (ibid.). He explains:

> This finality, experienced in both a prior and actual manner by *Dasein* in all the relations it encounters within-the-world, is an understanding *pro-posed* by the play of relations that make up the world, signification. The totality of these relations forms the significance that makes the meanings of words possible [. . .].
>
> (Stiegler, 1998, p. 250: emphasis added)

The word 'pro-posed' is key in unpacking this critique: Heidegger repeatedly attempts to present time as the result of the synthesis holding together retaining and awaiting, but he irretrievably gives emphasis to the latter; *Dasein* is always already *ahead-of-itself*, always already projected in a process of *in advance forming* that is itself determinately formed by the inescapability of death. *Dasein*'s past, however, is what is ahead of *Dasein*, constituting the *already-there* in which *Dasein* is forming its own *there*. This past is not merely *Dasein*'s individual past but the past of its community that *Dasein* inherits. This past is always already inscribed in *Dasein*'s individual and collective history, and according to Heidegger (2008):

> *With the existence of historical being-in-the-world, what is ready-to-hand and what is present-at-hand have already, in every case, been incorporated into the history of the world.* Equipment and work – for instance, books – have their 'fates'; buildings and institutions have their history. And even Nature is historical [. . .]. These entities within-the-world are historical as such, and their history does not signify something 'external' which merely accompanies the 'inner' history of the 'soul'. We call such beings "the world-historical".
>
> (pp. 440/388–889)

The past that is constitutive of the already-there refers to history not as an object to be observed, but rather as synthesis that has in any case contributed to the constitution of *Dasein*, prior to any possible separation between individual and communal histories. Similarly, the ready-to-hand is, as a historical process, containing and delivering the past. For this reason, Stiegler (1998) comments that the ready-to-hand as the world-historical, '[*weltgeschichtlich*] is a constitutive dimension of temporality, prior to and beyond the opposition between authentic temporality and intratemporality' (p. 18). A thing like a book is not the trace of the has-been, and it is certainly not a representation of history. It is rather history itself: it is forming the way my world presents itself to me; it is the conservation of memory and the inscription of knowledge through a process of 'externalisation'; indeed, it is time. When I read a book and perceive its meaning for the first time, I also perceive myself in the book through time: I make up my time through inscribed memories, not synthesised by me but necessarily resynthesised by me. The book, therefore, is a form of *pros-thesis*, which, according to Stiegler, means

> "placed-there-in-front." Pros-theticity is the being-already-there of the world, and also, consequently, the being-already-there of the past. *Pros-thesis* can be literally translated as pro-position. A prosthesis is what is proposed, placed in front, in advance; technics is what is placed before us [*la technique est ce qui nous est pro-posé*] (in an originary knowledge, a *mathésis* that "pro-poses" us things). Knowledge of mortality is knowledge of pro-position, but through these kinds of knowledge that are *tekhnai;* in a profound and diverse manner, it is the knowledge of a "primordial" de-fault: the de-fault of quality, of having to-be, destiny as predestination. The pro-position or technicity summons time.
>
> (1998, pp. 235–236)

Time is not possible without memory, that is, without a past, regardless whose past. In Stiegler's critique, memory becomes the matter of time. It is matter because of its exterior containment in tools, and it is also a matter of importance because of this containment's finite nature. Stiegler (1998) goes on to explain that technologies, like the book, are 'mnemotechnics' that contain the inscription of memories (p. 217). Therefore, he asserts that

> [a] tool is, before anything else, memory: if this were not the case, it could never function as a reference of significance. It is on the basis of the system of references and as a reference itself that I hear the "tool" that is "the creaking coach." The tool refers in principle to an already-there, to a fore-having of something that the *who* has not itself necessarily lived, but which comes under it [*qui lui sous-vient*] in its concern. This is the meaning of epiphylogenesis. A tool functions first as image-consciousness. This constitutivity of "tertiary memory" grounds the irreducible neutrality of the *who* – *its* programmaticality, including above all the grammar governing any language.
>
> (pp. 254–255)

Epiphylogenesis is a process of recapitulation during which a member of a community relies not solely on their own development but also on whatever has happened before them in order to individuate themselves. Whereas, epigenesis refers to the memory that an individual member of a species produces during its lifetime, namely, a memory that is subsequently lost with this being's death, epiphylogenesis is memory of the ancestors, exteriorised and retained in technologies inherited by the descendants. Through this 'structure of inheritance and transmission [. . .] the epigenetic layer of life, far from being lost with the living when it dies, conserves and sediments itself, passes itself down' (p. 140). In this respect, human history unfolds by means of the technological and the material schematisation of time. Every generation is thrown into a different technological *alreadyness*, and the coupling of the tool and the human being produces certain types of human beings, technological instruments and learning processes. In fact, Stiegler believes that epiphylogenesis is 'the very ideal of mathésis'[1] (p. 140).

What's more, and as underlined by Stiegler, this alreadyness is inscribed in both technology and language, namely, in processes conserving memory and defined by their finitude: they cannot contain everything, and *Dasein* cannot access everything or a lot at once. *Dasein* reactivates memory through technological and languaging implements and learns to be positioned towards a certain future. In other words, exterior memories are bound to be defined by selectivity while being themselves the 'criteria of selection' of *Dasein's* attention (Stiegler, 2011, p. 39).

The fact that Stiegler, contrary to Heidegger, understands technology and language as the two aspects of the same process is extremely useful for the critique articulated here, seeing that it opens up a space in which technology and language could be investigated so as to allow the emergence of a new philosophy of education, learning and teaching. Stiegler's critique, however, does not proceed to this direction – at least not through this perspective. Rather, it turns the question of exteriorisation into the question of reference and memory: technology and language appear to function on the premise that any one of their respective constitutive parts is a sign that refers to some other sign, whilst this process of reference ultimately constructs attention. At any moment, these systems rely on a selection of signs – that is to say, of dominant tertiary memories that condition anticipation and individuation. In effect, this perspective understands selectivity as the backbone of différance, making attention, as the selective process par excellence, the question of our time. I believe, however, that Stiegler's theorisation puts the process of schematisation at the margins of critique, indeed, at a time that media differentiation, remediation, and convergence, make the matter of schematisation quite pressing. Could we say, for example, that a book and a film about the Second World War are performing the same task, precisely because they are referring to the same event? What's more, can we infer that these media contain the same memories, or are we right to assume that the selectivity of memories is itself affected by the associating power of each respective medium? My answer is certainly in favour of the latter perspective, since I believe that the way tertiary memories are schematised, through their different media instantiations, and through our responses to them, are bound to play an important role in the way attention is formed. A further implication of this realisation would be the fact that attention does not unfold only referentially but also schematically. Perception is always already a type of synthesis and we thus need to investigate the synergetic synthesis that binds technology with language. Kant's famous example, namely, the one depicting the possibility of thinking time *as* time through spatialisation, points to the importance of metaphoricity that takes place through the medium. This process, which allows space to come close to time, does not take place in an abstract mental space of signs. Rather, these signs are possible through metaphorical exteriorisation that involves a hand utilizing tools, which are not simply memory. Indeed, my proposal here is that both Heidegger and Stiegler neglect the embodied and material schematising dimensions of metaphoricity, since Heidegger, on the one hand, identifies metaphor with metaphysics and technology, and Stiegler, on the other hand, neglects the material and embodied dimensions of schematisation – which are very much present in his earlier writings, while equating technology with memory and selectivity. As I have repeatedly argued in this book, a reframing of Heidegger's critique of technology would consist in addressing the processes that make any kind of Enframing (*Gestell*), *in advance forming*, revealing, and imagination (*Einbildungskraft*) possible. For the time being, though, I examine the ways in which Heidegger's prioritisation of time neglects to take into account the contribution of materiality and embodiment for the constitution of nearness.

4. *Dasein*'s embodiment

For a work that directs so much of its analytical power towards the seminal terms of the ready-to-*hand* and present-at-*hand*, *Being and Time (BT)* does not seem that much concerned with the actual human hand or the actual human body for that matter. Of course, on the one hand, this seems congruent with the expressed mission in *BT*, that is, to abandon and to destruct the metaphysical notions of form and matter, and, body and soul, which any metaphysical discussion would have addressed separately and as distinct entities. On the other hand, the body, as constitutive part of the phenomenon called *being-there* or *being-in-the-world*, ought to have gained more attention, even, if only, to claim its respective place opposite Heidegger's repetitive references to "soul". Indicative of this absence is David Krell's (1992) question:

> Did Heidegger simply fail to see the arm of the everyday body rising in order to hammer shingles onto the roof, did he overlook the quotidian gaze directed toward the ticking watch that overtakes both sun and moon, did he miss the body poised daily in its brazen car, a car equipped with turn signals fabricated by and for the hand and eye of man, did he neglect the human being capable day-in, day-out of moving its body and setting itself in motion? If so, what conclusion must we draw?
>
> (p. 52)

Heidegger's puzzlement by the body could be perhaps attributed to his intention to understand the human being in terms of existence, and which leads him to assert that *Dasein* 'does not measure off a stretch of space as a corporeal Thing which is present-at-hand', and that 'it does not "devour the kilometers"'(Heidegger, 2008, p. 140/106). In contrast, *Dasein* exists as a movement of 'bringing-close or de-severance' that is itself 'always a kind of concernful Being towards what is brought close and de-severed' (ibid.). In other words, *Dasein*'s spatiality is better understood in terms of nearness and, by extension, in terms of temporal orientation towards these beings that occupy its concerns. The following example is indicative, since Heidegger says that:

> One feels the touch of it [of the street] at every space as one walks; it is seemingly the closest and Realest of all that is ready-to-hand, and it slides itself, as it were, along certain portions of one's body-the soles of one's feet. And yet it is farther remote than the acquaintance whom one encounters 'on the street' at a 'remoteness' ["Entfernung"] of twenty paces when one is taking such a walk. Circumspective concern decides as to the closeness and farness of what is proximally ready-to-hand environmentally. Whatever this concern dwells alongside beforehand is what is closest, and this is what regulates our de-severances.
>
> (pp. 141–142/107)

Concern is the temporal orientation that decides as to the farness and the nearness of things. The tool, similarly to the body, withdraws so that these temporal engagements unfold. However, this withdrawal is never thought in terms other than privation. For this reason, even decades after *BT*, Heidegger (2001) writes:

> When the philosopher Thales, lost in thought, walked along a road, fell into a ditch, and was ridiculed by some servant girl, his body was in no way 'lost in space.' Rather, it was not present [. . .] when I am absorbed in something

'body and soul,' the body is not present. Yet, this 'absence' of the body is not nothing, but one of the most mysterious phenomena of privation.

(p. 85)

The body, as does the tool, withdraws into inconspicuousness after having established some kind of connection to the matter at hand. However, I believe that this absence need not be thought in terms of privation, since the body might as well be part of the very substratum that allows for the continual unfolding of any concern. Indeed, it may contribute to the synthesis of time *as* time. An example of what this bodily contribution would look like could have come from a potential exposition of *Dasein*'s directionality through the distinction between right and left. Heidegger thought that *Dasein*'s 'bodily nature has a whole problematic of its own', and I believe that orientation could have been a good entry point into the investigation of this problematic (ibid.). However, Heidegger, according to Dreyfus (1991), disconnects completely 'the issue of *Dasein's* embodiment from the issue of orientation' (p. 137). Indicative is the following discussion in which Heidegger (2008) argues that

> things which are ready-to-hand and used for the body – like gloves, for example, which are to move with the hands – must be given directionality towards right and left. A craftsman's tools, however, which are held in the hand are moved with it, do not share the hand's specifically 'manual' ["handliche"] movements. So although hammers are handled just as much with the hand as gloves are, there are no right- or left-handed hammers.
>
> (p. 143/109)

Heidegger does not believe that there is a difference between right-handed and left-handed people. Equally, he does not believe that there is a difference between these two directions, arguing that '[l]eft and right are not something "subjective" for which the subject has a feeling', and adding that '[b]y the mere feeling of a difference between my two sides I could never find my way about in a world' (ibid.). This, however, comes as a surprise, since Heidegger relies on Kant's essay *Concerning the Ultimate Ground of the Differentiation of Regions in Space* in order to discuss directionality, and, in this essay, Kant gives an account of the 'special bond' existing between the human body, direction and objects in space (Casey, 1997, p. 204). The discussion, according to Casey (1997), goes as follows:

> [T]hese material entities would be *unoriented*, lacking the definite directionality of "right" and "left," "up" and "down," "front" and "back." These paired terms, taken together, describe the three dimensions of space" and he [Kant] proposes a distinctively *corporeal deduction:* it is only because our own bodies are experienced as *already* bifurcated into paired sides and parts (e.g., right and left hands, chest and back, head and feet) that we can perceive sensible objects as placed and oriented in regions that rejoin and reflect our own bodily bifurcations. Things are not oriented in and by themselves;

they require our intervention to *become* oriented. Nor are they oriented by purely mental operation: the a priori of orientation belongs to the body, not to the mind.

(p. 205)

For Kant, the human body's structure is *transferred* onto things, and this movement of metaphor or transfer in space – as its Greek etymology suggests, is lived through the meaningful relations we share with things. The fact that the human being has two hands that point to two different directions, simultaneously exhibiting a possibility of the here-now, of the there-later, and also of the-here-there, could be thought as constitutive of *Dasein*'s specific spatiotemporality.

Thinkers, like cognitive linguists, Lakoff and Johnson (1980), do believe that human embodiment – as our foundational relation to the world – conditions not simply language but cognition itself. The case of 'orientational metaphors' is a good example of the way that time is constituted, since these metaphors bring together the familiar domain of embodied spatial experience of being-in-the-world and the intangible domain of time. Lakoff and Johnson believe that time *as* time first becomes possible through this transfer of time onto space. The human use of the binary *up* and *down* is telling of this hermeneutical process, which they name conceptual metaphor, allowing the perception of '*up*' as something positive and happy and the perception of '*down*' as something negative and sad. Expressions as the ones following make this clear: 'I'm feeling *up*! *That boosted* my spirits. My *spirits rose*. You're in *high* spirits. Thinking about her always gives me a *lift*. I'm *feeling down*. I'm *depressed*. He's really *low* these days. *I fell* into a depression. My spirits *sank*' (p. 15). Lakoff and Johnson argue that these phrases have a common 'physical basis', which is able to explain why '[d]rooping posture typically goes along with sadness and depression, [and] erect posture with a positive emotional state' (p. 15). In other words, Lakoff and Johnson reconceptualise metaphor in terms of a nearing process that allows not only for the logical proximity of conceptual systems but also for the constitution of cognition as hermeneutical synthesis performed by our *being-a-body-in-space*.

Maurice Merleau-Ponty's (1964) phenomenological distinction, between body-image and body-schema, similarly exemplifies the way that the human experience is mediated by the body. This contribution is, however, formulated not in terms of conceptual contribution, but rather in terms of schematisation. According to Merleau-Ponty:

We grasp external space through our bodily situation. A 'corporeal or postural schema' gives us at every moment a global, practical, and implicit notion of the relation between our body and things, of our hold of them. A system of possible movements, or "motor projects," radiates from us to our environment. Our body is not in space like things; it inhabits or haunts space. It applies itself to space like a hand to an instrument, and when we wish to move about we do not move the body as we move an object. We *transport* it without instruments as if by magic, since it is ours and because through it we have direct access to space. For us the body is much more than an instrument or a means; it is our expression in the world, the visible form of our intentions. Even our most secret affective movements, those most deeply tied to the humoral infrastructure, help to shape our perception of things.

(p. 5, emphasis added)

The use of instruments extends the body's ability to inhabit space just as the blind man's cane becomes 'an area of sensitivity, extending the scope and active radius of touch' (p. 143). It is not that the blind man touches that with which the cane interacts, but the cane's mediation allows for a certain connection to be established and for a certain spatiotemporality to unfold. The initiating power of the connection between the body-schema and the instrument makes further connections possible. What's more, the cane probably functions in the same way as the keyboard: through successive clicking, the user moves between places and rhythms that ultimately produce the time of the task at hand. In this way, we can think the absence of the body not in terms of privation but in terms of relating through and of transferring embodied relations. Merleau-Ponty's insight is very useful here: 'To get used to a hat, a car or a stick is to be transplanted into them, or conversely, to convert them into the bulk of our own body' (ibid.). The body as schema is itself a form of pre-phenomenological, pre-linguistic spatialisation that allows relatedness to and connectedness between its connecting parts and other things. As José Gil (1998) explains:

> The space of the body is made of plates, exfoliations, surfaces, and volumes that underpin the perception of things. These spaces 'contain' the relations of the body to things, insofar as they are integrated in the body itself and insofar as they are translated among themselves. The elaboration of these spaces in the course of the development of an infant's mobility and organic maturity implies their constitution as spaces decidable into other spaces – that is, their constitution thanks to their activity of the decoder-body, or the infralinguistic body, each of which is thus connectable to others, associating, intermingling, and boarding according to the laws of a specific mechanism.
> (cited in Hansen, 2006, p. 257)

Heidegger's specific understanding of metaphor and metaphysics, as inherently technological ways of meaning-production, does not allow for different conceptualisations of the body. On the contrary, language's specific revealing is set against technology's instrumentality and, in consequence, against the body's interaction with tools. For this reason, metaphor and language are dissociated from this interaction's contribution to the experience of closeness. For Heidegger, directionality is a matter of temporal familiarity, but the constitution of this familiarity is never set under investigation. Indeed, how do right and left come to be constituted as right and left in the first place? What is the effect of bodily directionality on thought, and how is thought to understand itself through the specific language of the body's bifurcation? Heidegger's (2008) answer invokes time once more, giving the following example:

> Suppose I step into a room which is familiar to me but dark, and which has been rearranged [*umgeraumt*] during my absence so that everything which used to be at my right is now at my left. If I am to orient myself the 'mere feeling of the difference' between my two sides will be of no help at all as long as I fail to apprehend some definite object 'whose position', as Kant remarks casually, 'I have in mind'. But what does this signify except that whenever this happens I necessarily orient myself both in and from my being already alongside a world which is 'familiar'? The equipmental context of a world must have been presented to *Dasein*. That I am already in a world is no less constitutive for the possibility of orientation than is the feeling for right and left.
>
> (p. 144/109)

In this respect, Heidegger (2008), just like Stiegler, interprets the tool in terms of memory, selectivity and finitude, arguing that orientation is rooted 'in the memory ["im Gedächtnis"]', which is in fact an allusion 'to the existentially constitutive state of Being-in-the-world' (p. 144/109). This account allows Heidegger to explain why *Dasein* cannot be involved with two projects at the same time, since, according to Heidegger there is always one futural possibility with which *Dasein* can be involved at any given time, and this, in consequence, suggests that temporal finitude determines being, orientation and nearness. This should not necessarily be the case, however, and Dreyfus (1991) argues accordingly that a fuller explanation is needed, explaining that for Heidegger

> not all equipment is accessible at the same time. I can turn to one thing or another but not both at once. These incompatible fields of action group simultaneously accessible things together in opposed regions called right/ left, and also front/back. But still without the body there could be no account of why there are just these regions. We would not be able to understand, for example, why the accessibility of right and left is not symmetrical, or why we must always 'face' things in order to cope with them. On Heidegger's account these would just remain unexplained asymmetries in the practical field. This is not inconsistent, but unsatisfying.
>
> (p. 137)

One great objection could be raised here: indeed, if the already-there is limitedly understood in terms of representation, memory and selectivity, in other words, if the already-there is understood strictly in temporal terms, then it is also assumed that the ready-to-hand is in any case already learned and already present. In this respect, absence; namely, the very thing that both Heidegger and Derrida attempt to foreground as constitutive of presence, is not revealed in its constituted dimension. However, there must be, always already, something absent, a lack of some sort, in order for *Dasein* to be able to inherit, recapitulate and receive anything. This absence is what allows the ready-to-hand to come near and to become ready in my hand in the first place. This absence cannot be thought simply in terms of a void that awaits to be filled by memories. It can also be thought as a spatiotemporal, corporeal, exterior and at times technological schema. In other words, this absence is a process of metaphoricity, bringing-close and synthesis that allows familiarity to be constituted as familiarity, the house to be constituted as house and the homely to be experienced as homely. Heidegger will never look into the constitution of this familiarity and this would have a great impact on his work.

5. Conclusion

Heidegger's early phenomenological descriptions of the ready-to-hand can be applied to any type of object, but what can this term reveal about the moment that the tool meets the hand? Why is even such a coupling possible? Is there some kind of learning involved in *Dasein*'s being accustomed to the ready-to-hand? Do the inconspicuousness of the tool and the withdrawal of the body suggest something beyond privation? Heidegger by no means raises these questions let alone responds to them. He only discusses the transition between the ready-to-hand and the present-at-hand in terms of privation. As already indicated, the very processes of learning to use and of producing a tool constitute in themselves important fields of inquiry, which are never investigated by Heidegger, since the temporal priority he ascribes to the tool's nature prevents him from doing so. In this light, familiarity, in whichever way it is conceived, either as alreadyness or simply as tool-structure, becomes essentialised and closed, and in this way the present-at-hand becomes the only possible means to access the ready-to-hand (see Dostal, 2006).

   This *aporia* intensifies during the later stages of Heidegger's career, that is, when Heidegger makes distinctions between tools like jugs and tools like airplanes. This distinction, however, is not supported by the existential analytic, since, according to Heidegger's specific phenomenological point of view, all tools are defined by their purposiveness; tools are used 'towards' something, 'in order to' do something and so as to bring things near. In this respect, technologies like nuclear factories and search engines are quite like hammers and shoes: they are defined by purposiveness, concern and nearness. These asymmetries, concerning the specificity of tools and the transition between their modes of being, might be what have led Heidegger (1972) to confess, later in his career, that '[t]he attempt in *Being and Time* [. . .] to derive human spatiality from time is untenable' (p. 23). Indeed, time and memory, as the singular elements of *Dasein*'s and technology's determination, cannot explain sufficiently the difference between older and modern technologies, authentic and inauthentic time. Heidegger's attempts to solidify *Dasein's already-there*, its forming power and indeed nearness through time and memory alone will intensify in the middle stage of his career. The catastrophic consequences of this theorisation will be discussed in the next chapter.

Note

1 *Mathésis* is the word for learning in Greek.

References

Arisaka, Y. (1995). On Heidegger's Theory of Space: A Critique of Dreyfus. *Inquiry*, *38*(4), 455–467.

Aristotle. (2012). *Physics: Book IV* (R. P. Hardie & R. K. Gaye, Trans.). In W. D. Ross (Ed.). Adelaide: The University of Adelaide Library.

Casey, S. E. (1997). *The Fate of Place, a Philosophical History*. Berkeley, CA: University of California Press.

Dostal, R. J. (2006). Time and Phenomenology in Husserl and Heidegger. In C. B. Guignon (Ed.), *The Cambridge Companion to Heidegger* (2nd ed., pp. 120–148). Cambridge: Cambridge University Press.

Dreyfus, H. (1991). *Being in the World: A Commentary on Division I of Heidegger's Being and Time*. Cambridge: MIT Press.

Hansen, B. N. M. (2006). *Bodies in code: Interfaces with Digital Media*. New York and London: Routledge.

Heidegger, M. (1969). *Discourse on Thinking* (J. Anderson & H. Freud, Trans.). New York, London, Toronto and Sydney: Harper & Row.

Heidegger, M. (1972). *On Time and Being* (J. Stambaugh, Trans.). New York and London: Harper & Row.

Heidegger, M. (2001). *Zollikon Seminars: Protocols-Conversations-Letters* (R. Askay & F. Mayr, Trans.). In M. Boss (Ed.). Evanston, IL: Northwestern University Press.

Heidegger, M. (2008). *Being and Time* (J. Macquarrie & E. Robinson, Trans.). Oxford: Blackwell.

Kisiel, T. (1993). *The Genesis of Heidegger's Being and Time*. Berkeley and Los Angeles, CA: University of California Press.

Krell, D. F. (1992). *Daimon Life: Heidegger and Life-Philosophy*. Bloomington, IN: Indiana Univeristy Press.

Lakoff, G., & Johnson, M. (1980). *Metaphors We Live By*. Chicago and London: University of Chicago Press.

Malpas, J. (2007). *Heidegger's Topology: Being, Place, World*. Cambridge, MA: MIT Press.

Merleau Ponty, M. (1964). *The Primacy of Perception: And Other Essays on Phenomenological Psychology*. Evanston, IL: Northwestern University Press.

Stiegler, B. (1998). *Technics and Time, 1: The Fault of Epimetheus* (R. Beardsworth & G. Collins, Trans.). Stanford, CA: Stanford University Press.

Stiegler, B. (2011). *Technics and Time, 3: Cinematic Time and the Question of Malaise* (S. Barker, Trans.). Stanford, CA: Stanford University Press.

5 Rootedness

Nearness as a political scheme

1. Introduction

The years following the publication of *Being and Time (BT)* have been characterised by Heidegger as the 'Turning' (*die Kehre*) of his thought, which can be understood, according to Sheehan (2001), as both the construction of the Turning and 'a shift in the way Heidegger formulated and presented his philosophy beginning in the 1930s' (p. 3). For Sheehan, the 'Turning' can also be thought as a shift from the 'the metaphysical ideal of being as full presence and intelligibility' to a 'radically inverted meaning of being grounded in finitude' (ibid.). In this context, Heidegger's focal question is not *what beings are or what Being is*, but 'what brings about being as the giveness or availability of entities?' (p. 7) This, subsequently, suggests that Heidegger's efforts to destruct metaphysics and to dismantle the centrality of the subject are accentuated. Indeed, Heidegger (2010) confesses in his *Letter on Humanism* that:

> The adequate execution and completion of this other thinking that abandons subjectivity is surely made more difficult by the fact that in the publication of *Being and Time* the third division of the first part, "Time and Being," was held back [. . .]. Here everything is reversed. [. . .] This turning is not a change of standpoint from *Being and Time*, but in it the thinking that was sought first arrives at the location of that dimension out of which *Being and Time* is experienced, that is to say, experienced from the fundamental experience of the oblivion of Being.
>
> (p. 157)

According to this confession, found in what Spiegelberg (1971) considers as Heidegger's most self-reflective text, the *Kehre* was intended to be a radical shift from the tradition that the philosopher aimed to de(con)struct but never successfully did, not at least, in *BT*. For this reason, the period following Heidegger's great work can be understood as his dialogue with his own thought. Indeed, later Heidegger's writings do not constitute 'a reorientation' of thought; they do not aim to replace 'one set of concepts by another' but attempt to shift 'the weight of emphasis from one term to another within his central distinctions' (Olafson,

1993, pp. 97, 98). This means that while early Heidegger attempted to reach the meaning of Being by questioning the entity that has Being as its problem, namely, *Dasein*, later Heidegger attempted to see *Dasein* through its fundamental relationship to Being. For this reason, Richard Wolin (1990) comments that, during this later period, 'when the term "*Dasein*" appears, it is frequently hyphenated to read "*Da-sein*" – Heidegger's way of indicating that the term is best understood as the "there" of "Being," rather than as an autonomously existent entity' (p. 131).

The shift from the meaning-making *Dasein* to the meaningful character of the world that *Dasein* inhabits is in accord with Heidegger's renewed emphasis on language's revealing role, which is supplemented with the human being's role as the listener of the world's Saying (Olafson, 1993). In this context, phenomenology's first-person perspective is considered inappropriate and is replaced by a form of language thought to be challenging propositional and technologised language.

By means of this new poetic and essentially metaphorical reformulation of his thinking, Heidegger's attention shifts from time and moves on to space indeed in a way that turns his Turning into a '(re)turning to Place' (Casey, 1997, p. 259). This emerging *topo-logy* is nevertheless devoid of philosophical rigor, constituting a strange amalgam of hermeneutic insights and of Nazi politics, and coinciding with Heidegger's allegiance to the Nazi movement that culminated in his rectorship at the Freiburg University in April 1933.[1] During this period, Heidegger thinks *nearness* according to the Nazi propaganda and with total disregard for technology's mediation. What's more, Heidegger contrasts technology to spirit (*Geist*) and paints the latter in such metaphysical overtones that makes it possible for scholars like Nicholas Tertulian and Jürgen Habermas to argue that 'beginning with the Rectoral Address of 1933, Heidegger's philosophy itself undergoes a fundamental transformation: it ceases to be a pristine "first philosophy"; it becomes [. . .] a veritable *Weltanschauung* or "world view"' (Wolin, 1990, p. 8). In the same vein, Jean-François Lyotard (1997) supports that during this period Heidegger's 'extraordinary thought has let itself be seduced in a very ordinary way by the tradition that always offers itself in the immediate context, "visible" for the world that succumbs to it. It is overcome by *Verfallenheit*'(p. 63). Finally, Sheehan (1988) characterises the Heidegger of this period as

> the dues-paying member of the NSDAP from 1933 to 1945 (card number 312589, Gau Baden); the outspoken propagandist for Hitler and the Nazi revolution who went on national radio to urge ratification of Hitler's withdrawal of Germany from the League of Nations; the rector of Freiburg University (April 1933 to April 1934), who told his students, "Let not theories and 'ideas' be the rules of your being. The *Führer* himself and he alone is German reality and its law, today and for the future [. . .].
>
> (p. 37)

This grim, but very accurate portrait of Heidegger, forces us at this point to wonder whether Heidegger's philosophy is inherently fascistic, or even ask, as Otto Pöggeler does: 'Was it not through a definite orientation of his thought that Heidegger fell – and not merely accidentally – into the proximity of National Socialism, without *ever truly emerging from this proximity?* (cited in Wolin, 1990, p. 1). This is not an easy question, and we can respond to it in many ways. There is, of course, the possibility that the philosophy Heidegger produced during this period is consistent with his earlier work. And, if this earlier philosophy is inherently fascistic then Heidegger's Nazi allegiance was, at least in part, consistent with it. If, however, Heidegger's philosophy is not fascistic, then the Nazi allegiance could represent Heidegger's betrayal of his own philosophy and possibly the manifestation of his opportunistic attempts for professional advancement. Yet, there could be a third possibility, namely, the possibility that Heidegger's politics is simply irrelevant to his philosophy, and that his political decisions constitute a misreading of his own work. Such a scenario, however, can neither exonerate the philosopher who trusted a nightmarish political ideology nor the philosophy that allowed the one formulating it to succumb to such a misreading. For this reason, Thomson (2005) infers that no approach can do justice to 'the philosophical integrity of Heidegger's thought' (p. 81). He also adds that

> the ongoing publication of his *Gesamtausgabe* (Complete Works) makes it increasingly obvious the Heidegger invoked his own philosophical views as justifications for his political decisions, and, as a result, even long-embattled Heideggerians are beginning to realize that a firm separation of Heidegger's politics from his philosophy is no longer tenable.
>
> (p. 81)

The recorded instances that present Heidegger's explicit saying that his political decisions were in agreement with his philosophy attest to that. For example, during a 1936 conversation, between Heidegger and his student Karl Löwith, the latter expressed his belief that 'his former mentor's "partisanship for National Socialism lay in the essence of his philosophy", and Heidegger responded that this was so '"without reservation, and added that his concept of "historicity" was the basis of his political "engagement"' (p. 75). However, if we remember here that it is precisely this notion that Stiegler (1998) criticised and in fact in terms of excluding technological mediation from the theorisation of the constitution of authentic time, we can easily see how this exclusion becomes political.

In the same vein, Thomson (2005) argues that it was Heidegger's concern with thinking that led to his political entanglement, asserting that the 'most immediate connection between his philosophy and politics' is Heidegger's 'long-developed philosophical vision for a radical reformation of the university' (p. 84). Indeed, Heidegger's conviction that technology and science impose restrictions on thinking and contribute to the compartmentalisation of knowledge led him to believe that the Nazi university policies would allow philosophy to regain its 'throne as the queen of the sciences' and determine a new way of thinking (p. 116).

Beyond any general criticism, it is quite pertinent to make here the following note: whatever Heidegger excluded from his philosophy of time and nearness, and, indeed, on the ground of its being dangerously entangled with metaphysis – that is, spatiality, corporeality, technology and metaphor, are the very elements of existence that could have rescued his philosophy from the Nazi propaganda. Indeed, the belief that Nazism would be the authentic and essential choice for *Dasein*'s future can be true only for a philosophy that prioritises the futural, the spiritual, and the immaterial determination of death. This prioritisation, along with an almost mystical return to the notion of the 'home', is accentuated during this period. Taking this into account, this chapter attempts to answer the following questions:

a How is technology understood in Heidegger, when the Turning is formulated and phenomenology is abandoned?
b How is nearness reconfigured through the reconceptualisations of time and space?
c How do *polis* and politics fit in this new theoretical framework?
d And, finally, how is Heidegger's thought formulated at this stage?

2. Technology in middle Heidegger

There are contradictory interpretations concerning Heidegger's account of technology in *BT*, speculating whether the powerful phenomenological descriptions of the ready-to-hand refer to an ontological ahistorical depiction of human everydayness or to a historical – and perhaps even politically skewed – analysis. Michael Zimmerman (1990) argues in favor of the latter, stating:

> *Being and Time's* phenomenological 'description' of everyday life was in part a negative political evaluation of industrial society. This political slant is partly responsible for the ambiguity present in *Being and Time's* account of everydayness. On the one hand, it purports to reveal the essential, timeless, or transcendental features of everyday life, on the other hand, those 'descriptions' are in some ways politically charged interpretations of everyday life in the specific historical circumstances of urban-industrial society. Early Heidegger's analysis did not make sufficiently clear the distinction between what were essential or universal features of everyday life and what were historical and specific aspects of it.
>
> (p. 17)

According to Zimmerman, Heidegger's expressed concern with the degradation of language – to the level of 'idle talk (*Gerede*)' – and of daily life was indicative of a general political anxiety concerning the future of Germany (p. 23). Following this position, Wolin (1990) argues that Germany's defeat in the Great War and the country's increased industrialisation created the fertile ground for different political responses concerning the envisioned future of the country and of technology's role in it. These responses, on the one hand, represented 'German neonationalist intellectuals', aspiring to create 'a *modern community*, one whose identity and structure would be forged through the utilization of the most advanced technological means in all spheres of life'[2] (pp. 79–80). On the other hand, these responses brought along a powerful anti-technological movement, which, according to Zimmerman (1990), was divided into two groups: the first group, the 'reactionary thinkers', felt 'threatened by the advances of modernity and industrial technology' and wished 'to defend the traditional ways of life that they believed were essential to the German identity' (p. 3). The other group, the '*völkisch* ideologues' lamented the endangered uniqueness of the German people (p. 21). They also bemoaned the

> spiritless mentality of modern economic society [. . .]. [This situation] called for a renewed contact with the natural and cosmic forces which, while inaccessible to the rational mind, were capable of rejuvenating and transforming the increasingly mechanized German spirit. According to *völkisch* ideologues these cosmic forces were at work in the common language, traditions, art, music, social customs, religion, blood, and soil which united a particular *Volk*.
>
> (p. 8)

Völkisch ideologues also asserted that the 'spiritual strength came from the rootedness in the natural soil of their homelands [. . .] and called for reconciliation with nature, not for the technological domination of it' (p. 9). In addition, they 'maintained that scientific rationalism, economic and political individualism, and industrial technology were behind such rootlessness' (ibid.). *BT* seems to echo much of this rhetoric, especially at the point at which Heidegger discusses *Dasein*'s individual authenticity in connection to being-with-Others (*Mitsein*), the destiny (*Geschick*) of the community and the destiny of one's people (*Volk*). During this discussion, Heidegger (2008) argues that

> if fateful *Dasein*, as being-in-the-world, exists essentially in being-with-Others, its historizing is a co-historizing and is determinative for it as *destiny* [*Geschick*]. This is how we designate the historizing of the community, of a people. Destiny is not something that puts itself together out of individual fates, any more than being-with-one-another can be conceived as the occurring together of several Subjects. Our fates have already been guided in advance, in our Being with one another in the same world and in our resoluteness for definite possibilities. Only in communication and in struggle does the power of destiny become free. *Dasein's* fateful destiny in and with its 'generation' goes to make up the full authentic historizing of *Dasein*.
>
> (p. 436/384)

It has been repeatedly stressed that the appropriation of the past can be theorised as a differential process during which *Dasein* chooses how to relate to its history and thus how to create its own time. In this case, however, Heidegger appears to believe in a deterministic communal destiny, which has in any case been decided without the thoughtful mediation of the individual. Heidegger does not explain how this pure determination of the communal (*völkisch*) destiny (*Geschick*) is to take place in an otherwise inauthentic and technologically formed social space, and thus his sharp distinction between authentic and inauthentic time, and, between language and technology, appears unattainable. What's more, Heidegger seems to assimilate here the possibility of authentic individuation to that of collective destiny, threatening in this way his own existential analytic that is based on autonomous choice. For this reason, Miguel de Beistegui (1998) argues:

> It is at this point of the analysis, toward the middle of the section, that the text, head on, blind to the consequences, precipitates itself, all too hastily, all too carelessly, in the abyss of steely and *völkisch* rhetoric. It will never quite recover from this journey.
>
> (p. 16)

The irreconcilable contradiction between *Dasein*'s most important possibility to be an authentic self, and its communal destined fate, indicates, as, I have already argued here, a greater problem in Heidegger's earlier understanding of technology. Indeed, in *BT*, both technology and language appear to be keepers of past memories that offer access to tradition. Each individual *Dasein*, however, needs to be free to choose the memories it wishes to singularly reappropriate. With the preceding passage, Heidegger seems to ignore the conditions of the possibility of inheritance, while distorting his own conception of existential choosing, synthesis and nearness. What's more, with the exclusion of corporeality, technology and metaphor, Heidegger will end up formulating a type of nearness that is deeply exclusionary ethnocentric and fascist and unquestionably very much focused on the notion of the 'home'.

2.1. A phenomenological interruption

As we have already discussed, *dwelling, being-there, being-in-the-world* and *nearness*, are predominantly time-related notions in early Heidegger. What's more, *Dasein's being in time* is discussed either as authentically unfolding through its orientation towards death or as inauthentically unfolding through its pragmatic ready-to-hand concerns. This important existential relation, however – between a human being and a tool – is completely erased during this middle period and is replaced by the abstract relation that connects a people (*Volk*) with its homeland, its destiny and spirit (*Geist*). Spirit was, of course, one of the notions that Heidegger was determined to avoid, as he considered the body/spirit-soul dichotomy to be part of the very metaphysical structure that he aimed to destroy. Leaving the body out of the existential analytic would have appeared to be consistent with Heidegger's efforts to destruct the oppositions that contain such a term, and a similar treatment would have been expected for the notion of spirit. This is not, however, the case. In contrast, when Heidegger (2008) attempts to think *Dasein*'s spatiality, we come upon the following passage:

> *Dasein's* spatiality [cannot] be interpreted as an imperfection which adheres to existence by reason of the fatal 'linkage of the spirit to a body.' On the

contrary, because *Dasein* is 'spiritual', *and only because of this*, it can be spatial in a way which remains essentially impossible for any extended corporeal Thing.

(p. 419/368)

This suggests, however, that it is because Dasein 'is "spiritual" (this time in quotation marks, of course) that it is spatial and that its spatiality remains original. It is by virtue of this "spirituality" that *Dasein* is a being of space and, Heidegger even underlines it, only by virtue of such a "spirituality"' (Derrida, 1989, p. 31). Equating spirituality with spatiality, impacts greatly the Heideggerian existential analytic: indeed, if *Dasein*'s spatiality is grounded in spirit, then being-in-the-world, which is itself possible via the ready-to-hand, is to be conceptualised either as spirit or as technology – the latter being predominantly understood as the lack of spirit. Derrida (1989) seems to assert that Heidegger is moving towards such an opposition; one between spirit and time, on the one hand, and space and technology, on the other, through the very priority he ascribes to questioning (*Fragen*). Questioning is for Heidegger what opens up the path of thinking. *Dasein*'s spatiality as such is then differentiated from other types of corporeal spatialities – like the animal's, for example, precisely because *Dasein's spatiality* is understood through *Dasein*'s potentiality to be a spiritual and a questioning being, or to put this otherwise, through *Dasein's* potentiality to be a being questioning Being. Taking into account this distinction, technology is to be found in the realm of the animalistic: technology, similarly to the animal, is devoid of spirit and is preventing *Dasein* from questioning its own being. Technology de-spiritualises the world. It is no surprise then that during this middle period, Heidegger chooses to talk about spirit in order to speak about *nearness* and chooses to speak about the *home* in order to oppose technology. De Beistegui (1998) explains the necessity of this route:

> To economy proper – that of labor and of the Worker, that of production in the age of technology – Heidegger wished to oppose or liberate another economy, which he never acknowledged as such, not even as an economy: it is the law of the *oikos*, of the home and the hearth, the law of the proper, of the national and the native. This law is not the effect of labor and production, but that of (the) work (of art) and the poet. It arises out of the necessity to articulate the *Da* of *Sein*, to delimit the space or the place of being. In other words, it is an economy of being [. . .].
>
> (p. 158)

The turning towards the home could have served as the basis for an insightful phenomenological account, finally describing the daily lives of beings that have bodies, that live in houses and learn to be in the world by means of the spatiotemporality and familiarity of the home. These beings would first experience nearness in that which offers them the possibility of being at all. This is not, however, the case. Instead, Heidegger's re-turning to the home serves his aesthetic nationalism and allows him to articulate an abstract future-oriented 'spiritual' discourse of nearness.

2.2. Nearness as spirit

Articulating a rhetoric heavily influenced by Nazi ideology and reactionary thinking – intermingled with some of his previous phenomenological and existential insights – Heidegger's Rectoral Address (1985), *The Self-Assertion of the German University*, given in 1933, urges students and teachers of the university to join him in their common 'spiritual mission', asserting that:

> The following of teachers and students awakens and grows strong only from a true and joint rootedness in the essence of the German university. This essence, however, gains clarity, rank, and power only when first off all and at all times the leaders are themselves led – led by the unyielding spiritual mission that forces the fate of the German people to bare the stamp of its history.
>
> (p. 470)

As Derrida (1989) notes, the comeback of the spirit is striking now: 'In the wings, spirit was waiting for its moment. And here it makes its appearance. It presents itself. Spirit itself, spirit in its spirit and in its letter, *Geist* affirms itself through the self-affirmation of the German university' (p. 31). The return of the spirit does not come without consequences: in contrast, it demands a radical revision of the discourse of authenticity, assuming that each individual *Dasein*'s potential is overtaken by the *Volk*'s destiny. What's more, the call of consciousness, which needs to be singularly heard and synthesised by each individual *Dasein*, is now irretrievably replaced by the collective listening to the call of the leader, whilst phenomenological awareness (the *letting-be* relationship) is transformed into a conformist attitude of compliance. Derrida comments that the

> [s]elf-affirmation wants to be (we must emphasize this wanting) the affirmation of spirit through *Führung*. This is a spiritual conducting, of course, but the *Führer*, the guide – here the Rector – says he can only lead if he is himself led by the inflexibility of an order, the rigor or even the directive rigidity of a mission (*Auftrag*). This is also, already, spiritual. Consequently, conducted from guide to guide, the self-affirmation of the German university will be possible only through those who lead, while themselves being led, directors directed by the affirmation of this spiritual mission.
>
> (p. 32)

The deferment of personal responsibility constitutes Heidegger's first betrayal of phenomenology. Indeed, his recommendation that students should comply with the new Student Law, which 'sought to organise students according to the *Führerprinzip* in an effort to integrate the universities into the National Socialist state'[3], attests to the great distance stretching between this new discourse of spirit and the 'spirit' of *BT*'s discourse (Heidegger, 1985, p. 475). With his speech, Heidegger also welcomed the banishment of 'academic freedom', which he saw as 'freedom from concern, arbitrariness of intentions and inclinations, lack of restraint in what was done and left undone' (p. 476). He thus concluded that via the banishment of academic freedom: '[t]he concept of the freedom of the German student is now brought back to its truth. Henceforth the bond and service of the German student will unfold from this truth' (p. 476).

Heidegger's second betrayal of phenomenology comes with his affirmation of the Nietzschean will-to-power, that is, the sheer 'decisionism' that, according to Zimmerman (1990), 'resonated not only with Jünger's writings, but also with those of the leading Nazi jurist, Carl Schmitt, and the ideologue Alfred Rosenberg' that 'scorned the Weimar weaklings who claimed pitifully "that they wanted the best, only circumstances were too difficult to achieve it. But they forget or do not want to admit that they did not really *will* it"' (p. 72). Likewise, Heidegger (1985) argued that

> [o]nly a spiritual world gives the people the assurance of greatness. For it necessitates that the constant decision between the will to greatness and a letting things happen that means decline, will be the law presiding over the march that our people has begun into its future history.
>
> (p. 475)

Even though, we need to assume that the '*letting-things-happen*', which Heidegger denounces here, does not refer to the letting relationship, but rather to the *laissez-faire* state of the Weimar Germany, we cannot help our astonishment before the ease with which these phenomenological connotations escape Heidegger's attention. Conversely, the struggle for national determination seemed perfectly suited with his vision of the university's reformation, which he saw possible through science's liberation from technologised thinking. Science was to become the most radical form of questioning: an originary mode of thinking that went back to the inception of Greek thought. Whilst this discussion of originary thinking returns with a vengeance in the *Introduction to Metaphysics* (2000), we can say almost with certainty that at the time of the Rectoral Address, Heidegger conceives this thinking as the instrument destined to deliver Germany to its greatness. In any case, the phenomenological aspiration to receive what is yet unthought and to say what is unsaid is abandoned through these authoritative declarations. Later, in his 1945 retrospective essay, *The Rectorate 1933/34 – Facts and Thoughts*, Heidegger (1985) will argue that his original speech underlined '[t]he essence of truth as the letting be of what is, as it is' (p. 487). However, the structure and actual wording of the address do not seem to allow for such a reading. On the contrary, the speech distorts the most important connotations of the letting relationship, especially so when Heidegger turns his attention towards the teachers' role, stating:

> If we will this essence of science, the body of teachers of this university must really step forward into the most dangerous post, threatened by constant

uncertainty about the world. If it holds this ground, that is to say, if from such steadfastness – in essential nearness to the hard-pressing insistence of all things – arises a common questioning and a communally tuned saying, then it will gain the strength to lead.

(p. 475)

Questioning – that is, the basic determination of the human being as the one belonging to the spiritual realm – is now understood as nearness to all things. This nearness, instead of being the uncertain experience of receiving the alien through the familiar, is now transformed into 'a communally tuned saying'. In other words, the philosopher-leader cannot maintain his thought's phenomenological awareness at this critical historical time; the world can be given neither time to speak nor space to wonder. The philosopher must speak in its place, and thus all others voices must be silenced. Otherness – like technology, for example, cannot speak: it cannot participate in nearness. On the contrary, it needs to be reduced to a dangerous threat and to be fought as such. At the same time, however, students are encouraged to join the *Labor Service*, the *Military Service* and the *Knowledge Service* that, according to Zimmerman (1990), 'correspond to the divisions of the ideal city in Plato's *Republic*' (p. 68). In view of this similarity, Wolin (1990) detects a correspondence between Heidegger's vision and Jünger's directives concerning the creation of the soldier-worker. He argues:

> The inordinate emphasis in the speech on the virtues of labor and military service betrays the pronounced influence of Jünger's doctrines. And thus, taking his cue from *Der Arbeiter*, Heidegger decides that if the society of the future will be composed of worker-soldiers and soldiers-workers, then the universities, too, must do their part by producing *student-worker-soldiers*.
>
> (pp. 88–89)

In summary, during this period, Heidegger seems to adopt both substantive and instrumental interpretations of technology: technology is the threat to any true form of thinking, but at the same time technology is the fundamental tool that turns students into soldiers and workers. Therefore, Zimmerman's (1990) comment is quite accurate, stating that: 'Paradoxically, Heidegger believed, moving beyond nihilism and violence brought by modern technology was possible only on the condition that humanity first submit to the claim of modern technology' (p. 47). It thus becomes increasingly clear that Heidegger's understanding of technology – at least, during this period – is inconsistent to say the least. In the next section, I elaborate further on this.

2.3. *Rootedness and technology*

As already noted, there are still considerable doubts concerning the theorisation of technology in *BT*. During this middle period, however, Heidegger's position, concerning modern technology's negative and dehumanising effects, is solidified. Later, in the *Introduction to Metaphysics* (1953/2000), he will write:

> This Europe, in its unholy blindness always on the point of cutting its own throat, lies today in the great pincers between Russia on the one side and America on the other. Russia and America seen metaphysically, are both the same: the same hopeless frenzy of unchained technology and of the rootless organisation of the average man. When the farthest corner of the globe has been conquered technologically and can be exploited economically; when any incident you like, in any place you like, at any time you like, becomes accessible as fast as you like; when you can simultaneously 'experience' an assassination attempt against a king in France and a symphony concert in Tokyo; when time is nothing but speed, instantaneity, and simultaneity, and time as history has vanished from all *Dasein* of all people; when the tallies of millions at mass meetings are a triumph; then, yes then, there still looms like a specter all this uproar the question: what for? – where to? – and what then?
>
> (p. 40/29)

Rootlessness, considered as modern technology's worst impact, legitimises Heidegger (1985) to replace *nearness* with *rootedness*, which, for him, expresses accurately 'the power that mostly preserves the people's strengths, which are tied to earth and blood' (p. 475). Of course, certain other displacements take place so that this new conceptual organisation will unfold: First, space, which was marginalised in Heidegger's earlier thought, is now defining his ideological agenda, and, secondly, the alien – that is, the unfamiliar and the uncanny, thought to be an essential condition for authenticity, *alētheia* and true nearness, is now treated as a menace. Dreyfus (1989) says:

> Directly contradicting his early emphasis on man's essential experience of not being at home, later Heidegger strives to give us "a vision of new rootedness which someday might even be fit to recapture the old and now rapidly disappearing rootedness in a changed form."
>
> (p. 75)

In the text, *Why Do I Stay in the Provinces?* (1934),[4] Heidegger (2003) underscores the 'inner relationship of [. . . his] own work to the Black Forest and its people [which] comes from a centuries-long and irreplaceable rootedness in the Alemannian-Swabian soil' (p. 17). This relationship is not in any case discussed phenomenologically or through *Dasein*'s situated existence in connection to its environment. On the contrary, Heidegger resorts to a melodramatic discourse, discussing the connection binding his 'work-world' with the region of peasant life, the purity of which stands in stark contrast to the 'the world of the city [which] runs the risk of falling into a destructive error' (pp. 16, 18). Heidegger writes of this world:

In large cities one can easily be as lonely as almost nowhere else. But one can never be in solitude there. Solitude has the peculiar and original power not of isolating us but of projecting our whole existence out into the vast nearness of the presence [*Wesen*] of all things.

(p. 17)

For Heidegger, the technologically produced public space of the modern city constitutes a kind of modern homelessness that is juxtaposed to the rootedness found in the *Heimat* (homeland). In order to defend this position, Heidegger attempts to bring to the fore the inherent spirituality of the German land by way of Hölderlin's hymns *Germania* and *The Rhine*. His respective lectures on the hymns were, in fact, delivered a few months after his resignation from the rectorship, but Philippe Lacoue-Labarthe (1990) argues that 'his "Hölderlinian" preaching is the continuation and prolongation of the philosophico-political discourse of 1933' (p. 12). In a similar vein, Kathleen Wright (1994) identifies Heidegger's distorted hermeneutic tactics and the

four 'textual strategies' that Heidegger employs in his first lecture course on Hölderlin's late hymns, which serve, she contends, as a means of endowing the hymns with a proto-Nazi gloss. These 'strategies' are (1) 'to fragment the textual unity of 'Germania' by reading into it fragments drawn from Hölderlin's letters and poems'; (2) 'to alter the tone or mood of the poem'; (3) 'to disregard the tropology of the poem both by denying that Hölderlin maintains a distinction between the figurative/fictional and the literal/factual and by reversing the meaning of images,' and (4) 'to regender Germania by substituting a masculine for a feminine voice.'

(cited in Nowell Smith, 2013, p. 157)

In this light, Heidegger's return to space coincides with the distortion of the very process that his thinking attempts to become: the poetic image. Instead of letting poetic imagery to be freely received, Heidegger imposes interpretations that temporalise space and emphasise the historicality of the homeland. This historicality offers its people the possibility of seizing a nationalistically envisioned future and, thus Heidegger asserts that '[e]arth and homeland are understood in a historical sense' (cited in Elden, 2002, p. 38). The historicising of the land, which previously excluded the spatial, the social and technological conditions producing it, dominates the discussion in *Introduction to Metaphysics*, namely, the text that allows Heidegger to rethink the Greek *polis*, the political sphere and the organisation of the nation. De Beistegui (1998) explains that Heidegger's argument there

can be seen to be engaged in a double gesture: on the one hand, he thinks the possibility of a use of the national that would be free from nationalism as well as from the form of the nation-state; on the other hand, he re-evaluates this latter and distinctively modern form of political organization, that is the

nation-state – this very state which in effect is the vehicle and the most effective servant of technology – by way of a reflection on its forgotten essence,
namely the *polis* [. . .].

(p. 114)

In the next section, I turn to Heidegger's text with the intention to show what
happens to the description of the *polis*, when the technological dimensions that
constitute it are excluded.

3. A re-turn to the *polis*

After having forcefully criticised the failures of the historical organisation of the
nation state, Heidegger attempts to rethink the political realm from its origin,
that is, from the perspective of the ancient Greek *polis* (De Beistegui, 1998). As
a result, Heidegger (2000) proceeds to a problematisation of the term, arguing
that

> [o]ne translates *polis* as state (*Staat*) and city-state (*Stadtstaat*); this does
> not capture the entire sense. Rather, *polis* is the name of the site (*Stätte*),
> the Here, within which and as which Being-here is historically. The *polis* is
> the site of history, the Here, *in* which, *out of* which and *for* which history
> happens. To this site of history belong the gods, the temples, the priests, the
> celebrations, the games, the poets, the thinkers [. . .].
>
> (pp. 162–163/117)

After conceiving space in terms of historicity, Heidegger proceeds to equate the
political with the historical – and the ontological realm, and it is precisely this
identification that, according to Lacoue-Labarthe (1990), is responsible for Heidegger's ideological misconceptions. He writes:

> It is clear that, for Heidegger, 'political', in the sense in which he became
> politically committed, means 'historical' and that the act of 1933, having
> regard to the University, but also, beyond it, to Germany and to Europe,
> is an act of foundation or re-foundation. And is no less clear that in 1933
> National Socialism embodied that historical possibility or at least it was
> bearer of it.
>
> (p. 17)

Polis, as an essentially historical possibility, does not seem to require the involvement or the singular interpretations of particular human beings who become
citizens. Rather, *polis* necessitates the unfolding of pure non-technologically mediated time. According to Heidegger (2000), the ancient Greek city instantiated
the essence of the *polis*, and this connection represented a unique possibility for
the German people whose language was considered to be the natural heir of the
ancient Greek language. Heidegger strengthened this connection by means of

his return to Sophocles's *Antigone* and through his particular reading that excludes technology from the theorisation of the Greek city. In what follows, I focus on a few indicative examples of this reading, derived from Heidegger's discussion of the following chorus lines:

a   Line 334 reads: 'πολλὰ τὰ δεινὰ κοὐδὲν ἀνθρώπου δεινότερον πέλει', and is translated as: 'Wonders are many, and none is more wonderful than man'.[5] The line discusses the extraordinariness of the human being. This unique-ness of the human being is further expounded through the chorus' examples praising human achievements, such as fishing, hunting and agriculture. Hei-degger, however, argues that '[t]hese [. . .] notions from cultural anthropol-ogy and the psychology of primitives' present the 'inception of history' as 'primitive and backward'. For him, the inception of humanity is the most 'uncanny (*unheimlich*) and mightiest', and it needs to be thought as 'mys-tery' (pp. 165–166/119). Heidegger's own translation here is especially suggestive of his intentions: by choosing to translate the δεινόν not as won-der but as *unheimlich* (unhomely), Heidegger underlines that the essence of the human being lies in the transcendence of familiarity and its limits. He thus detects the necessity of violence and explains that '[h]umanity is violence-doing' (p. 172/123).

b   Line 358 describes the human being as 'παντοπόρος· ἄπορος'. Heidegger takes this phrase as a pair of words that go together, even though they are separated by a semicolon. He subsequently concentrates on the meaning of *poros* as a kind of passing or route and relates it to the previous line (line 334) in order to suggest that this phrase comes to define the '*deinotaton*'. For him, this suggests that humanity that instantiates the uncanny opens paths into 'all domains of being' (p. 162/116). With this deeply meta-physical interpretation, however, Heidegger overlooks the necessary and the pragmatic technological constitution of path-making and also the standard translation of *pantoporos* as resourceful, bearing the connotations of the technological mediation that allows the traversing of space and the creation of new paths.

c   The next pair of words that Heidegger discusses is 'ὑψίπολις· ἄπολις' (line 370), which again is separated by a semicolon, suggesting that *ipsipolis* (high-citied, the citizen who respects the law and belongs to the city) refers to the explanation that precedes this word, and *apolis* (the one who does not respect the laws and brings disgrace to the city) describes what follows. Obviously Creon, the king of the city and the keeper of its laws, is, for the chorus, the *ipsipolis*, and Antigone, namely, the one defying the laws in order to bury her dead brother's body against Creon's wishes, is the *apolis*, namely, the one who is convicted to be buried alive, in a cave, which is probably outside of the city. The opposition here is thus between the historical moral law of the state and the eternal law of the gods. The two appear to be in conflict. Heidegger, however, sees an important connection in this pair of words. First, he explains that when *polis* is translated as city state, this does not fully explain what an ancient Greek *polis* is, since, '(t)he *polis* is the site of history, the Here, *in* which, *out of* which and *for* which history happens' (p. 162/117). In this respect the poets, the thinkers, the priests and the rulers are the 'violence-doers' and the 'creators' who by

> rising high in the site of history are also *apolis*, without city and site, lonesome, un-canny, with no way out amidst beings as a whole, and at the same time without ordinance and limit, without structure and fittingness (*Fug*), because they *as* creators must first ground all this in each case.
>
> (p. 163/117)

Of course, the point worth making here is that it is precisely not the creator of the laws and the head of the state – that is, Creon – who becomes *apolis*. To the contrary, the one who becomes *apolis* is Antigone, whose decisions change her personal history and the history of the city. In fact, it is only with Antigone's suicide that it finally becomes clear who was in the wrong and who was in the right. In this respect, it is not the creator but the citizen who obeys the eternal laws of gods that becomes *apolis*. It could then be argued that the two terms cannot refer to the same being: the poet, the creator, or the philosopher cannot be both ὑψίπολις and ἄπολις because the *polis* is the site of history and, in consequence, the one who falls out of the *polis* falls out of history. Indeed, it appears that Heidegger's discussion is contradictory: the violence-doer needs to fall out of history in order to transform the city as the site of history. Casey (1997) comments:

> But if that is the case, this person also breaks with place – breaks away from place and breaks place itself. This is tantamount to leaving the *polis* and to destroying it as a 'place of history.' Heidegger does not hesitate to draw this consequence, contrary as it is to his earlier praise of the place of the *polis* as a scene of constructive activity.
>
> (p. 263)

It is then evident that Heidegger's analysis is highly problematic: the transfiguration of the city into pure time cannot allow space for individual and collective becoming, namely, the very processes that take place in and as the city. Casey argues that this dead-end has profound effects on Heidegger's own becoming, adding that: '[i]nspired by his allegiance to a Nazi ideology of violence, Heidegger himself, the creative thinker, has here fallen into [. . .] "confusion"' (p. 264). In other words, '[t]he creative action undoes its own basis: the limit. By becoming undelimited, it ceases to have a place in which to *be* creative' (ibid.). Antigone's own individuation is a testimony to the necessity of polyphony, dispersal and differentiation in space. But this is only possible if the *polis*, contrary to Heidegger's interpretation, is not conceived as '*polos*', as 'pole' and 'vortex' – gathering everything around the centre, but as that which allows dispersal, temporal deferral and spatial differentiation (1984, cited in Elden, 2000). This possibility can be traced, according to Casey (1997), back to *Dasein*'s spatiality, commenting that:

> If the transcendental condition of dissemination is bodily thrownness, the transcendental condition of multiplicity is spatiality. For only in the spread-outness of spatiality can *Dasein* disseminate itself into the multiplicity of 'beings which it is not.' The manyness and otherness of these beings – their being outside *Dasein* and their being next to each other – require a laid out-spatiality that answers to, even as it connects deeply with, the bestrewed bodiliness of *Dasein*.
>
> (p. 260)

This kind of multifaceted spatiotemporality is constituted as the retention of the past (through the technological realm), as the thrownness in the present and as the multiplicity of projections thrown in the future. Time cannot be a single destiny, and time cannot but be mediated by technology (Stiegler, 2003). Indeed, *Dasein*'s thrownness in an already-there is constituted through technologically inscribed retentions in space. In this respect, Antigone is able to appropriate her own choices and to choose her own heroes by means of her access to the laws of gods, as they are inscribed in temples, in social practices and documents. Heidegger (1968) ignores the technological constitution of the city and even downgrades the role of writing by considering Socrates, namely, the one who did not write, as the purest thinker of the West.[6] Bernard Stiegler (2003), in contrast, points to technology's constitutive role for the emergence of both the Greek *polis* and philosophy, arguing that, '[i]t is [. . .] mnemotechnics that makes possible the writing of laws, the founding of cities, the construction of geometric reasoning [. . .] the practice of philosophy' (p. 154). He then adds that technics

> appears well *before* Plato, and appears first of all as the *question of transformation and becoming* [*devenir*] (raised by the economic crisis associated with the development of navigation, money, and thousands of other new technics that appear at that time) in the Greek cities [. . .]. And it is not simply a question of technics, but also and above all of *mnemo*technics, that is to say of technics of the future, in its capacity profoundly to transform the conditions of being together, the terms of the law, the rules of life, etc.
>
> (p. 155)

The neglect of technology's constitutive role, for the co-constitution of the *polis* and for the emergence of nearness, results to the construction of new dichotomies; indeed, ones between non-technologically mediated nearness and technologically mediated rootlessness – that is, what Heidegger would later call distanceless-ness. For this reason, Heidegger (2000) supports that the creative violence of the human being should be expressed in terms of art (*technē*), instead of technology, since the former constitutes a mode of 'knowing' that 'means neither art nor skill'

(p. 169/122).[7] He also asserts that *technē* and nature (*phusis*) refer to the same mode of knowing, since

> [*p*]*husis* means the emerging sway, and the enduring over which it thoroughly holds sway. This emerging, abiding sway includes both 'becoming' as well as 'Being' in the narrower sense of fixed continuity. *Phusis* is the event of *standing forth*, arising from the concealed and thus enabling the concealed to take its stand for the first time.
>
> (pp. 15–16/11–12)

The juxtaposition of *technē* – as nature – to technology – as something entirely different from both art and nature – consolidates Heidegger's conviction, that there can be a type of *polis* and a type of political formation that is not mediated by technology but is instead rooted in the promise of the German land and the inheritance of its people. He, therefore, blatantly ignores the very conditions that allow this inheritance to exist and also to become – through the very texts he studies – his own means of individuation. As Stiegler (2003) comments, 'this terrifying political outcome is made possible *precisely* because Heidegger does not raise the question *here* of the actual conditions of this inheritance, inasmuch as they are already inscribed in its original technicity' (p. 158). The question of inheritance is the question of origin. Heidegger (2000) alludes to such a discussion when addressing the origin of the human being or even when considering language as the one producing the human being. This flickering thought, however, is not drawn to its conclusions. Such a route could have potentially revealed a process of mutual invention between language and humanity, during which the one doing the inventing and the one being invented are inseparable, rendering, in this way, the notion of origin untenable (see Stiegler, 1998). Stiegler (1998) moving precisely into this direction, proceeds to a deconstructive understanding of technology, seeing the human being as that being that lacks an origin and that exists through the prosthesis of both technology and language. The anthropological evidence, which Heidegger so readily dismisses, proves, according to Stiegler, the strong interconnection existing between, on the one hand, the skeletal and the brain development of the human being and, on the other, the simultaneous use of tools and language. In this respect, technology and language construct, invent and transform the human being – that is, they allow a process of *exteriorisation* that transforms an interior that does not exist before this exteriorisation. It is, in this way, that the technological realm constitutes time as the space of différance, rooted in technics. Stiegler (2003) writes:

> It is a temporality within which a living being, in particular the one that we call man, is constituted in relation to the temporality of a technics which is itself a technical development or becoming, which is the main dimension of becoming for human beings.
>
> (p. 156)

Heidegger understands technology's formative role for the givenness of inauthentic time. This axiological movement will remain in the background of his thought and will be repeatedly instantiated in his efforts to describe a type of time-synthesis, image or nearness that exists outside of technics. However, this juxtaposition, between technological and non-technological time, cannot be sustained on the grounds of his analytic of technology, since, as I have already pointed out, any piece of equipment, even a social networking site or an app, can be understood through the schema of purposiveness, projection and nearness. However, the presence/representation dichotomy cannot be useful either, considering that it is unable to explain the distinct hues of modern technological instances and to account for their propensity to produce distancelessness. What lies then at the heart of the distinction, between older and modern technology, is the way these technologies schematise time and instantiate nearness.

4. Conclusion

In middle Heidegger, the phenomenological study of nearness is suspended in favor of an ambiguous notion of nearness configured in terms of spirit, of rootedness and homeland. In this respect, the experience of being-at-home, which is coupled in early Heidegger with the undifferentiated *they-self*, is now deemed essential for solitude, authentic temporality and nearness. In the 'pure' nature of the German homeland, Heidegger appears to have found – even though he never says so explicitly – an 'origin' that is not mediated by technology. Heidegger makes a similar discovery, when he locates a creative type of thinking in the originary political experience of the Greek *polis*. All these discoveries come, however, at a great cost. First, the notion of the 'home', even though heavily relied upon, is understood one-dimensionally: de Beistegui (1998), in fact, wonders if this connection has harmed Heidegger's thought, asking:

> Are the *topoi* bound to be thought as *oikoi*? Not necessarily. I would want to suggest that if the spatiality of *Dasein* as being-in-the-world does indeed presuppose a certain mode of dwelling, it does not necessarily imply that mode of dwelling that became central to Heidegger's thought in the 1930s, and to which the political engagement of 1933–4, as well as the subsequent confrontation with this engagement, remained indebted: the national and the native, the *Heimatliche* and the *Heimische*.
>
> (p. 159)

According to Emmanuel Levinas (1990), the homeland, as theorised by Heidegger, becomes a lot more dangerous than technology – namely, the very thing that Heidegger considered as the most frightening threat – since it tends to gather by excluding otherness. He comments:

> One's implementation in a landscape, one's attachment to *Place*, without which the universe would become insignificant and would scarcely exist, is

the very splitting of humanity into natives and strangers. And in this light technology is less dangerous than the spirits *[genies]* of the *Place*.

(p. 232)

For all these reasons, Leslie Paul Thiele (1995) comments that Heidegger's returning to the house could have easily been a returning to the household instead of the homeland. In fact, the unexplored embodied familiarity of the house, and its important role for the opening up of the possibility of everydayness, becomes the screaming absence in Heidegger's existential analytic: this intimate aspect of instrumentality and the kind of experiences people have when gathered in the house, resting, bathing, daydreaming, eating, sleeping or tending to each other's needs are marginalised before the home's spiritual standing.[8] This complete break with the phenomenology of the everyday is perhaps what prevents Heidegger from seeing the tragedy of his own historical time and also allows him to compare Auschwitz to the food industry. Heidegger (1949) says:

> Agriculture is now a motorized food industry, the same thing in its essence as the production of corpses in the gas chambers and the extermination camps, the same thing as blockades and the reduction of countries to famine, the same thing as the manufacture of hydrogen bombs.
>
> (cited in Lacoue-Labarthe, 1990, p. 34)

Lacoue-Labarthe claims that this analysis proves Heidegger's denial to attend to the things taking place around him and to pay heed to the phenomenon itself, asserting that Heidegger 'refused to admit that it was ultimately the duty of thought to confront that particular phenomenon and to seek to take responsibility for it' (p. 33). He adds:

> the extermination of the Jews (and its programming in the framework of a 'final solution') is a phenomenon which follows *essentially* no logic (political, economic, social, military etc.) other than a spiritual one, degraded as it may be, and therefore a historical one. In the Auschwitz apocalypse, it was nothing less than the West, in its essence, that revealed itself. And it is thinking that event that Heidegger failed to do.
>
> (p. 35)

In this respect, Heidegger's political mistakes cannot be separated from the errors of his thought. With the ending of this period, however, Heidegger begins to think time and nearness beyond the spiritual and the futural realms. Indeed, with *The Origin of The Work of Art* (originally published 1936), Heidegger (1975) sees history unfolding as essentially the strife between earth – that stands for the concealed realm of being – and world – that stands for revealing and bringing to light. Through this new negotiation of powers, Heidegger finds a way to talk about nearness, noting that that earth and sky might appear contradictory but, in truth, instantiate the letting relationship. This relationship was from the beginning described as perception – both receptive and active, reiterating the Kantian imagination. At this point, however, Heidegger attempts to separate this type of new thinking from imagination, stating the following:

> If we fix our vision on the essence of the work and its relation to the happening of the truth of beings, it becomes questionable whether the essence of poetry [*Dichtung*], and this means at the same time the essence of projection, can be sufficiently thought from imagination.
>
> (cited in Sallis, 1990)

The delimitation that Heidegger draws between poetic image, projection and imagination does not, however, dismiss imagination but the usual renderings of imagination as flight and fancy. For this reason, Sallis (1990) comments that:

> [Y]et it is Heidegger, perhaps, most of all who has provided the means for surpassing such impoverished concepts of imagination, especially through his interpretation of the Kantian transcendental imagination, which radicalizes imagination to the point where it merges with *Dasein* itself. If one notes, too, that the entire discussion of poetry and of projection in 'The Origin of the Work of Art' serves essentially to elaborate the opening of the *Da*, the opening of the space of truth, then it is doubly surprising that Heidegger, in effect, sets imagination aside. Even more so, if one considers the possibility that imagination, sufficiently deconstructed, would seem eminently fit to name that peculiar *active reception* that Heidegger has shown to characterize artistic creation.
>
> (pp. 185–186)

In the next chapter, I turn to Heidegger's attempt to describe a new way of thinking that embodies the letting-relationship and that, despite, his own intentions, involves imagination as time-synthesis, as nearness and metaphoricity.

Notes

1 Heidegger resigned from this position in 1934.
2 Ernest Jünger, namely, one of the greatest influences on Heidegger's thought, belonged to this group.
3 Translator's comment: footnote 8 in Martin Heidegger (1985, p. 475).
4 In this text, published as a newspaper article, Heidegger explains why he rejected two invitations (one in 1929 and another in 1933) to assume professorship at the Humboldt University of Berlin.
5 See Sophocles (1891).
6 He discusses this point in his essay *What is Called Thinking?* (1968).
7 For this reason, he translates ($\pi\acute{o}\rho o\varsigma$) as path, instead of resource. He also translates ($\mu\eta\chi\alpha\nu\alpha\tilde{\iota}\varsigma$) as arts, instead of machines.
8 Albert Borgmann (1984) offers such a discussion of the house, and especially the stove, in an attempt to make a clear distinction between a thing and a device. This discussion is very much in debt to later Heidegger's accounts of things thought

either as gatherings of the fourfold or as meaningless resources (*Bestand*). In any case, Borgmann's account points to a possible direction of this discussion, writing that:

> A thing, in the sense in which I want to use the word here, is inseparable from its context, namely, its world, and from our commerce with the thing and its world, namely, engagement. The experience of the thing is always and also a bodily and social engagement with the thing's world. In calling forth a manifold engagement, a thing necessarily provides more than one commodity. Thus a stove used to furnish more than mere warmth. It was a focus, a hearth, a place that gathered the work and leisure of a family and gave the house a center. Its coldness marked the morning, and the spreading of its warmth the beginning of the day. It assigned to the different family members tasks that defined their place in the household. [. . .] It provided for the entire family a regular and bodily engagement with the rhythm of the seasons that was woven together of the threat of cold and the solace of warmth, the smell of the wood smoke, the exertion of sawing and of carrying, the teaching of skills, and the fidelity to daily tasks. These features of physical engagement and of family relations are only first indications of the full dimensions of a thing's world (pp. 41–42).

References

Borgmann, A. (1984). *Technology and the Character of Contemporary Life: A Philosophical Inquiry*. Chicago and London: Chicago University Press.

Casey, S. E. (1997). *The Fate of Place, a Philosophical History*. Berkeley, CA: University of California Press.

De Beistegui, M. (1998). *Heidegger and the political: Dystopias*. London: Routledge.

Derrida, J. (1989). *Of Spirit: Heidegger and the Question* (G. Bennington & R. Bowlby, Trans.). London: University of Chicago Press.

Dreyfus, H. (1989). Beyond Hermeneutics: Interpretation in Later Heidegger and Recent Foucault. In G. Shapiro & A. Sica (Eds.), *Hermeneutics: Questions and Prospects* (pp. 66–83). Amherst: University of Massachusetts Press.

Elden, S. (2000). Rethinking the Polis, Implications of Heidegger's Questioning the Political. *Political Geography, 19*(4), 407–422.

Elden, S. (2002). *Mapping the Present: Heidegger, Foucault and the Project of a Spatial History*. London: Continuum.

Heidegger, M. (1968). *What Is Called Thinking?* (F. D. Wieck & J. G. Gray, Trans.). New York: Harper and Row.

Heidegger, M. (1975). The Origin of the Work of Art (A. Hofstadter, Trans.). In *Poetry, Language, Thought*. New York: Harper and Row.

Heidegger, M. (1985). The Self-Assertion of the German University and The Rectorate 1933/34: Facts and Thoughts. *Review of Metaphysics, 38*(3), 467–502.

Heidegger, M. (2000). *Introduction to Metaphysics* (G. Fried & R. Polt, Trans.). New Haven, CT: Yale University Press.

Heidegger, M. (2003). Why Do I Stay in the Provinces? (1934). In M. Stansen (Ed.), *Martin Heidegger: Philosophical and Political Writings* (pp. 16–18). London: The Continuum International Publishing Group Inc.

Heidegger, M. (2008). *Being and Time* (J. Macquarrie & E. Robinson, Trans.). Oxford: Blackwell.

Heidegger, M. (2010). *Basic Writings: From Being and Time (1927) to The Task of Thinking (1964)*. In D. F. Krell (Ed.). London and New York: Routledge Classics.

Lacoue-Labarthe, P. (1990). *Heidegger, Art and Politics: The Fiction of the Political* (C. Turner, Trans.). Oxford Blackwell.

Levinas, E. (1990). Heidegger, Gagarin and Us (S. Hand, Trans.). In *Difficult Freedom: Essays on Judaism* (pp. 231–234). Baltimore: The Johns Hopkins University Press.

Lyotard, J.-F. (1997). *Heidegger and 'the Jews'* (A. Michel & M. S. Roberts, Trans.). Minneapolis, MI: University of Minnesota Press.

Nowell Smith, D. (2013). *Sounding/Silence: Martin Heidegger at the Limits of Poetics.* New York, NY: Fordham University Press.

Olafson, F. (1993). The Unity in Heidegger's Thought. In C. B. Guignon (Ed.), *The Cambridge Companion to Heidegger* (pp. 97–121). Cambridge: Cambridge University Press.

Sallis, J. (1990). *Echoes: After Heidegger.* Bloomington, IN: Indiana University Press.

Sheehan, T. (1988). Heidegger and the Nazis. *The New York Review of Books, 35*(10), 38–47.

Sheehan, T. (2001). A Paradigm Shift in Heidegger Research. *Continental Philosophy Review, 32*(2), 1–20.

Sophocles. (1891). *Antigone* (R. Jebb, Trans.). In R. Jebb (Ed.). Cambridge: Cambridge University Press.

Spiegelberg, H. (1971). *The Phenomenological Movement: A Historical Introduction.* The Hague: Nijhoff.

Stiegler, B. (1998). *Technics and Time, 1: The Fault of Epimetheus* (R. Beardsworth & G. Collins, Trans.). Stanford, CA: Stanford University Press.

Stiegler, B. (2003). Technics of Decision an Interview. *Angelaki, 8*(2), 151–168.

Thiele, L. P. (1995). *Timely Meditations: Martin Heidegger and Postmodern Politics.* Princeton, NJ: Princeton University Press.

Thomson, I. D. (2005). *Heidegger on Ontotheology: Technology and the Politics of Education.* New York: Cambridge University Press.

Wolin, R. (1990). *The Politics of Being: The Political thought of Martin Heidegger.* New York: Columbia University Press.

Zimmerman, E. M. (1990). *Heidegger's Confrontation with Modernity: Technology, Politics, and Art.* Bloomington, IN: Indiana University Press.

6 Re-turning home
Nearness in later Heidegger

1. Introduction

In his later writings, Heidegger understands nearness in connection to non-subjectivist receptive poetic thinking – thought to be threatened by modern technology. This opposition clearly suggests that thinking is susceptible to technology's influence, indicating that there is a kind of communication between the two processes. The nature of this communication is of the utmost importance for this book, and by illuminating it, I will be able to discuss modern technology's characteristics and respective interactions with thinking, imagination and nearness. However, it is not easy to rely on Heidegger's later critique of technology in order to discuss imagination in any way, since there appears to be a sharp distinction between Heidegger's earlier first-person phenomenological approach, which is understood in connection to imagination, and his later poetic mystical path, thought to have nothing to do with individual and possibly imaginative thinking. In this respect, any attempt to trace imagination's relation to technology in the later texts would appear futile. Indeed, according to Heideggerian theorists, like Dreyfus (1989), Heidegger's methodology, during this later period, departs from the phenomenological hermeneutics of hiddenness and enters into a process that 'consists in taking some particular scientific achievement, political issue, or ritual as a case of "truth setting itself to work" and putting into words what this paradigm means for the practices it brings into focus' (p. 82). In this way, the link between individual imagining and technological manifestation is lost. Richard Palmer (1984), however, argues that there is a continuity between Heidegger's two types of investigation, and that 'Dreyfus [is] pushing Heidegger into denying things he did not explicitly deny but kept a resolute silence about' (p. 85).

In *A Dialogue on Language*, we can catch a glimpse of Heidegger's (1982a) own beliefs concerning the continuity of his thought and the supposed abandonment of terms like *phenomenology* and *hermeneutics* in his later work. In this text, Heidegger explains the motive behind the change of terms, asserting that '[t]hat was done, not – as is often thought – in order to deny the significance of phenomenology, but in order to abandon my own path of thinking to namelessness' (p. 29). He then adds that hermeneutics 'derives from the Greek *hermeneuein*' that takes us back to the god Hermes, 'the divine messenger' acting as the one who 'brings the message of destiny', and thus '*hermeneuein* is that exposition which brings tidings because it can listen to a message' (ibid.). He also explains that it was this 'original sense' of hermeneutics that prompted him 'to use it in defining the phenomenological thinking that opened the way to *Being and Time*', allowing him to think the ontological difference that he later came to call the 'hermeneutic relation' (p. 30). This assertion suggests accordingly that Heidegger does not abandon his earlier phenomenological insights, but rather intensifies the phenomenological/hermeneutic perspective through a different route. As Palmer (1984) comments: Heidegger, during this period,

> simply pursues the hermeneutical without naming it, becomes the text-interpretive philosopher par excellence, meditates on the "eventing" of language (the central problem in hermeneutics), and the hermeneutical placement of man in the world (without calling it hermeneutics). The fact that he did not, after *Being and Time*, turn away from his guiding question of the *meaning* of being and how being *discloses* itself to man (even though he abandoned a transcendentally oriented way of interrogating being) suggests that Heidegger is becoming *more* not *less* hermeneutical, since hermeneutics is centered on the process of being grasped by *meaning* in an event of disclosure. He abandons transcendental modes of thought but he does not abandon the essence of the hermeneutical.
>
> (p. 90)

The fact that Heidegger's later hermeneutics is more concerned with the 'hermeneutic relation' and less with the hermeneutical being, suggests, according to Palmer, that Heidegger invites us 'to try to think the essence of interpretation from out of the phenomenon of the hermeneutical' (p. 91). In this respect, Heidegger never 'gets definitively "beyond" meditating on the interpretation process that constitutes the existence of man' (ibid.). Interpretation coincides with Being and is thought as the process that delivers the human being into the nearness of things, corresponding, in this way, to the process of perception that early Heidegger attempted to sketch via the Kantian and the Aristotelian interpretations of imagination. Taking this knowledge into account, I do not so much support the idea that Heidegger remains a hermeneutical phenomenologist to the end – if that is, he ever was one – but rather that Heidegger's later investigations should not be opposed to his earlier ones: one should, instead, consider this thinking as the philosopher's journey towards thought through phenomeno*logy* – that is, through language as λόγος, which was earlier defined by Heidegger as the voice of imagination (*logos* is, after all, φωνή μετά φαντασίας). In this respect, what changes in Heidegger's later writings is not so much the topic of investigation as the route that constitutes now, more than ever, a questioning of *nearness* in and through language. Ziarek (1994) comments:

> This peculiar hermeneutics does not therefore present an interpretive task; it does not 'read' the meaning or truth of Being but instead draws attention to

the proximity, the nearness, inscribed in what Heidegger calls *Entsprechung* –
a fold into the way language occurs and 'speaks' [. . .].

(p. 9)

Following a similar path, I strongly maintain two points: first, that one of Heidegger's enduring themes, even though abandoned as terminology, is imagination. Imagination, as presented in previous chapters, is intrinsic to interpretation,
incorporating an element of spontaneity and creativity and an element of passivity
and acceptance as instantiated in the letting-relationship. This does not mean that
the focus of this investigation remains the same: the displacement of the subject
signals the displacement of time and, therefore, my second point is this: While
language is presented as the basic process that *lets things be* and affords nearness, Heidegger's own metaphorical language says much more about the way
nearness and the poetic realm unfold than his explicit argumentative language.
What's more, his discussion concerning the work of art, as a site for truth, emphasises the spatiotemporal dimensions of revealing and accounts for the material and
embodied aspects of its unfolding. This in turn provides us with an opportunity to
reconsider poetic image as a mode of presencing that does not belong to language
exclusively. Heidegger's (1969) discussion of a new type of thinking in the form
of the letting-things-be relationship, the *'releasement toward things'* or *'releasement toward things'* or *Gelassenheit* underscores this realisation, disclosing that this
type of thinking is possible by means of *Dasein's* embodied and spatiotemporal
situatedness coupled with technology (p. 54). By addressing this metaphorical-
hermeneutic thinking, it finally becomes possible to discuss how the ready-to-
hand becomes the ready-to-hand as familiar prosthetic exterior and learnt object.
For all these reasons, I turn now to Heidegger's later discussions on metaphor.

2. Returning to the house (of being)

The *Letter on Humanism* constitutes a recapitulation of several of Heidegger's
(2010) important ideas and also an indirect response to the Sartrean defense of
Cartesian subjectivity. In this essay, Heidegger reiterates his position concerning
human existence, which he considers as the *site of Being* or the *clearing of truth*.
In an attempt to illustrate the way through which the human being participates in
this unveiling, Heidegger discusses the relation that beings share with language,
with thinking and Being, asserting that:

> Thinking accomplishes the relation of Being to the essence of man. It does
> not make or cause the relation. Thinking brings this relation to Being solely
> as something handed over to it from Being. Such offering consists in the fact
> that in thinking Being comes to language. *Language is the house of Being.* In
> its home man dwells.
>
> (p. 147: emphasis added)

'Language is the house of Being'. *Dasein* dwells in a house that does not belong
to it, and it is not built by it, but it is inadvertently its house. It is perhaps a house

found; a house always already granted to *Dasein*; a house given and inherited. *Dasein* dwells in this house, and dwelling constitutes this house as a house. In fact; can there be a house without some sort of dwelling always already taking place in and through this house? Can *taking place* even take place without dwelling taking place? What would be the place of such taking place, and what will be the object of such taking? What is the meaning of Heidegger's metaphor? What is the meaning of Heidegger's language and of language in general? These are the questions guiding this discussion, which is focused on one of Heidegger's most enigmatic constructions; namely, a construction that comes too near to Heidegger's notion of *nearness* and *poetic image*. An understanding of the image depicting *language as the house of being*, and of the text that contains it, will possibly allow us to see the types of images that language and technology afford.

In the *Letter on Humanism*, Heidegger, says that technology goes back to Greek '*technē* as a mode of *alētheuein*, a mode, that is, of rendering beings manifest [Offenbarmachen]' and thus constitutes 'a form of truth' (p. 166). However, as we have previously seen, Heidegger does not place technology and language on the same level, whilst their opposition is reflected in the dichotomies of *alētheia* and *orthotis*, of authenticity and inauthenticity, and in the one between language-art-originary image and technology-metaphysics-representative image. Both sides of these dichotomies constitute ways that allow beings to manifest themselves out of absence, but one of these sides is clearly considered more proper than the other. Metaphor, as repeatedly stated by Heidegger, is undoubtedly on the side of the metaphysical, the technological and the derivative. An understanding then of the 'house of being', this distinctly metaphoricophilosophical conception, could open up some remaining *aporias* in Heidegger's thought and inform us about the kind of thinking Heidegger is attempting to sketch, the way he reconceptualises metaphor – if at all – and the lens through which he perceives language and technology as forms of emergence. Lastly, the investigation of this metaphorical language can also shed some light on the way Heidegger reconceptualises language as *nearness* and *dwelling*, which he then juxtaposes to modern technology as *distancelessness* and *homelessness*.

With these points in mind, we return to the house of Being. So what of the 'house of Being'? Heidegger sketches here a structure – the house, after all, is both literally and metaphorically a structure, which is not constructed by the human being. Rather, the human being's becoming unfolds through the dwelling of the house, whilst the human being's thinking coincides with this dwelling. The human being dwells in the house as it receives it as house, and it thinks thinking as it receives it as thinking. The human being's role in all this, Heidegger (2010) reminds us, is defined by its 'ecstatic existence [. . .] as "care"', which has always already been disclosed as the synthesis of time, as imaginative projection towards the future, and as interpretation that receives things *as* things (ibid.). In other words, the human being participates in the as-structure of dwelling, but it is not itself the originator of this structure; the origin rather comes from the possible future as a repeatable past and the possible past as a livable future. In this respect, the human being's origin comes from absence itself. The human being does not produce thinking or language. Rather, thinking and language produce the human by letting themselves be articulated by the human being; by letting the human being come near and dwell in them and as them.

But what does precisely thinking do? '[I]n thinking Being comes to Language', which is its home, explains Heidegger (p. 147). Still, what is the meaning of this homecoming? Certainly, language is not Being and neither is thinking. So what is thinking? Is it the place constituted by Being's coming to language? Heidegger comments, that '[t]hinking [. . .] lets itself be claimed by Being', and that '[t]hinking accomplishes this letting' (pp. 147–148). The letting relationship comes to the fore once again, pointing to a process that has in any case, always already, begun by receiving. Reception is a form of schematisation, since, in thinking, Being can be schematised and indeed be imagined as the one who speaks; as the one who is and dwells in a house. As stated in the later essay *A Dialogue on Language* (1982a): '[w]hen imagination wells up as the wellspring of thinking, it seems to me to gather rather than to stray. Kant already had an intimation of something of the sort' (p. 48). Imagination, as a form of gathering and synthesis, is revisited once again: imagination allows for thinking. However, the home of Being is that state in which thinking can say something, and therefore language is indissolubly bound to this thinking. Still, language is not the origin of Being. Language is a home in need of destruction: language truly speaks, when it breaks down, and it is, for this reason, that '[t]hose who think and those who create with words are the guardians of this home' (Heidegger, 2010, p. 148). They free language from ordinary use and imagine it through 'poetic creation' (ibid.). Thinking is not derivative of language and neither is language of thinking. As Heidegger (1982a) asserts, in a commentary of the 'house of Being', found in *A Dialogue on Language*, '[e]ven the phrase "house of being" does not provide a concept of the nature of language' (p. 22). It rather 'touches upon the nature of language without doing it injury' (ibid.). Indeed, 'the essential *being* of language cannot be anything linguistic', and as such we can infer that the phrase, the house of Being, is not linguistic either (p. 23). What is at stake then, is the 'manifestation of Being', and we need to keep in mind that any utterance is not readily a site of this manifestation. Besides, language is not essentially linguistic (Heidegger, 2010, p. 151). Language, Heidegger (2010) says, can constitute the true presencing of things, but it can also be 'an instrument of domination' that calculates and orders beings (ibid.). Therefore, Heidegger claims that a more essential aspect of language, one lying closer to the 'nearness of Being', is manifested through the 'nameless' (ibid.). The nameless *is* that which is not *denominated* but, nevertheless, *predicated*. The nameless *is*. Being *is*. Even though *Being* is denominated, it is mostly defined by predication here. Predication through the adjective avoids 'the metaphysics of presence' and escapes substantialism, denoting the presence of the alien in the familiar. What's more, predication takes us back to the copula of the verb *is*. And, this is Heidegger's question all along. Heidegger comments that we say 'there is/it gives ["es gibt"] Being' (p. 162). So what is at stake is the element corresponding to this *there is/it gives*, which is, however, absent. Better yet, it constitutes an absence giving presence and allowing for the nearing of that which is near. For this reason, Heidegger declares that '[t]he nearness occurs essentially as language itself'

(p. 161). Being is what is near, not as fully known and certainly not as completely unfamiliar. For this reason, Heidegger explains that 'thinking in its saying merely brings the unspoken word of Being to language' (p. 179). Being constitutes both a realisation and a promise, and to that end Heidegger writes that '[t]hinking builds upon the house of Being', warning us at the same time, that '[t]he talk of the house of Being is no transfer of the image "house" to Being. But one day we will, by thinking the essence of Being in a way appropriate to its matter, more readily be able to think what "house" and to "dwell" are' (p. 177).

The realisation is that Being is always already near. The promise is that one day we will be able to truly think that which is near and to think what the house and dwelling are. However, Heidegger (1982a) is careful to quickly dismiss any assumptions concerning his own thinking, namely, the thinking attempting to think thinking and to do so metaphorically, explaining that his thinking is neither metaphorical nor metaphysical, and it 'must not be taken as a mere hasty image which helps us in imagining what we will, such as: house is a shelter erected earlier somewhere or other, in which Being, like a portable object, can be stored away' (ap. 26). Such an explanation, however, is not in any way satisfactory, since nothing that is able to dwell as the human being dwells can be thought as a portable object. The human relation to the house is not a relationship of *in-one-another-ness* but always already a metaphorical relationship of nearness and dwelling, and this relation is precisely the one from which Heidegger's phenomenology shies away, probably because the experience of living in a house, as that which is most familiar and in consequence most elusive, can never be a mere sensuous, and easily representable, temporal source, which offers stock for potential transferences of meaning. Living in a house is always already an unfamiliar multifaceted terrain, with both phenomenal content resisting theorisation and phenomenological content demanding new approaches to thinking. Despite the rich theorising potential of the house, early Heidegger, as we know, left it unexplored while making *dwelling, being-in* and *being-near* his ways of translating a basic spatiotemporal existence into a temporal one, erasing thus several aspects of Being and most importantly erasing space. With Heidegger's new metaphor, things are different: Heidegger finally sees that neither the house belongs to the sensuous realm nor Being belongs to the intelligible one. Being, after all, can never be fully present: it is rather always already in the process of coming into presence through and as this metaphor's unfolding. In other words, Heidegger's metaphor encapsulates the meaning of nearness: it is that for which we await but is never fully present. Being does not become a concept or a name; it constitutes the eternal postponement of meaning and the interruption of meaning as presence. Therefore, Heidegger (1982a) argues that his metaphor 'differ[s] from sings and chiffres, all of which have their habitat in metaphysics' (p. 26). It is rather a 'hint', a 'gesture' or a 'bearing' (ibid., pp. 24, 26). Derrida (2007) gives a similar account of this metaphor, commenting that:

> One might be tempted to formalize this rhetorical inversion where, in the trope 'house of Being,' Being tells us more or *promises* us more about the house than the house does about Being. But this would be to miss what

the Heideggerian text means to say in this place, to miss what is, if you will, most proper to it. Through the inversion we're considering, Being has not become the proper essence of this supposedly known, familiar, nearby being, which is what one believed the house to be in the common metaphor. And if the house has become a bit *unheimlich*, this is, not because it has been replaced by "Being" in the role of what is nearest. We are therefore no longer dealing with metaphor in the usual sense or with a simple inversion permutating places in a usual tropical structure.

(p. 69)

It would appear then that Heidegger's metaphor, or what Derrida calls quasi-metaphor,[1] could be Heidegger's originary image, namely, a type of thinking that is not propositional, representational or representative; it is neither theoretical nor practical nor sensible nor non-sensible. It rather *dwells on* (in) the house. These metaphors do not substitute one word – or concept – for another, as the traditional understanding of metaphor assumes, but rather challenge the very notions of concept and of metaphor, allowing what is most familiar to become unfamiliar and what is most homelike to become unhomely (*unheimlich*). Being comes near through the breakdown of meaning and in its twofoldedness: It is through this breakdown that Being illuminates beings and is itself illuminated as the process that *lets beings come near*. This is, however, one of metaphor's basic features; indeed, a feature of any metaphor, and this is perhaps why Derrida argues that metaphor is in a way constitutive of Being. We can only speak of Being metaphorically, while every metaphor of Being tends to allow the sight of Being but also hide it through this process' withdrawal. That the withdrawal is understood by Heidegger as metaphysics proves that Heidegger's impression was right: metaphor lies at the heart of metaphysics. The withdrawal, however, also proves him wrong: Being cannot be thought in distinction from metaphor, which is itself a much more multifaceted process than the model of metaphysical opposition that supposedly constructs it. Derrida (2007) comments:

> Being being nothing, not being a being, it cannot be expressed or named *more metaphorico*. And therefore it does not have, in the context of the dominant metaphysical usage of the word "metaphor," a proper or literal meaning that could be intended metaphorically by metaphysics. Consequently, we can no more speak metaphorically on its subject, than we can properly or literally. We will always speak of it only *quasi*-metaphorically, according to a metaphor of metaphor, with the overload of a supplementary trait, a re-trait. A supplementary fold of metaphor articulates this retreat/retracing, repeating the intra-metaphysical metaphor by displacing it, that is, the very metaphor that the withdrawal of Being has made possible.

(pp. 65–66)

Derrida offers his own metaphor about metaphor or the withdrawal of metaphor or even the withdrawal of Being. For Derrida, there is always a 'retrait' of metaphor that functions as Heidegger's focal metaphor of the '*house of Being*', since

[i]t is not at all the case that I am starting out from a word or a known or determinate meaning (*retrait*) to think about Being or metaphor; rather I will come to comprehend, understand, read, think, allow the withdrawal in general to manifest itself only if I begin with the withdrawal of Being as a withdrawal/redrawing of metaphor in all the polysemous *and* desseminal potential of the *retrait*.

(2007, p. 167)

Derrida's point does not, however, resolve the problem. Rather, it suspends it, since what is at stake is precisely this trembling uncertain ground that allows the possibility of thinking metaphor or of thinking Being in terms of the re-trait. The re-trait works as a trait: it is not an origin but the trace of a process having in any case always already begun. It is the hermeneutic circle, namely, a reveal-ing, which can never reveal an entirety but can at least promise a certain degree of nearness to what is revealed through the revealing of the true nature of near-ness. Derrida (2007) sees his own and Heidegger's metaphors – the one about the house of being, and others that describe the *neighbourliness* shared by poetry and thinking – as processes destroying the metaphysical conception of metaphor, explaining that '[s]uch a catastrophe therefore inverts the metaphoric trajectory at the moment when, having overflown all borders, metaphoricity no longer allows itself to be contained in its so-called metaphysical concept' (p. 68).

In the light of this reconceptualisation of metaphor, I propose to discuss metaphoricity as a process inherent in hermeneutics, as constitutive of the herme-neutic circle and indicative of the workings of imagination. Non-metaphor, quasi-metaphor, or perhaps metaphoricity's own unfolding seems to implicate at least four terms from Heidegger's focal metaphor: the house, language, thinking and Being. Heidegger invites us to think these terms together and through the multifaceted ways that they reflect each other. Put in different terms: Heidegger invites us to think of the various ways these notions draw correspondences between each other. This process summons the human being and Being, and the significance of this sum-moning will be shortly explained. However, let us, first, turn to the two metaphors that I have already discussed here – that is, the one proposing that '[t]hinking builds upon the house of Being', and the other assuming that '[l]anguage is the house of Being', and trace the analogical correspondences that construct Heidegger's phi-losophy of metaphor and his philosophy of nearness that invites us to think:

a   Thinking as the being that builds a house;
b   Being as the being that dwells in a house;
c   Language as the house;
d   Language as the being built upon by thinking.

Now, if we look carefully at these analogies, it is evident that there are even more analogies running underneath, carrying more or less the following meanings:

a   Thinking, Being and Language are thought as beings that perform thinking, being and languaging.

b   Thinking, Being and Language are also thought as beings in general, namely, beings that exist *as* beings.
c   This process summons the human being as that being that participates in thinking, Being, and in language.
d   This process summons the human being as that being that participates in these processes while attempting to think their nature as beings, through its own mode of being, in and through them.

The question is one of endless correspondences, allowing us to rearticulate the question of the categorial intuition and even move beyond analogy. There are no strict analogies between these terms, but there appears to be a relational ground that allows different domains to come together. The human being is constitutive of this ground and at the same time the one attempting to think it. In this respect, these distinct domains do not stand over against each other; they are rather enfolded in each other. What's more, Being is present in all beings, but it is also the mediator for whatever is to be thought, while the human being participates in this thinking. This realisation leads Stellardi (2000) and Gasché (1997) to clearly see analogy at the heart of metaphoricity and to see metaphoricity at the heart of Being. In fact, Gasché (1997) asserts:

> Since analogy is, according to Aristotle, not only one genre of metaphor but the metaphor par excellence in so far as it is based on an equality of relations, the doctrine of the analogy of being – whatever the meaning of analogy may be – indicates that a certain metaphoricity is constitutive of the very unity of being. The as-structure of understanding unearthed by Heidegger characterizes understanding and the saying of Being as hinging on a movement of transfer.
>
> (p. 302)

Stellardi (2000) goes deeper into this, commenting that

> [t]he possibility of the analogy must in all cases preexist, albeit not always explicitly, as a differential space of language, or rather in the general text of the inscribing of all traces, of all forms. A trait capable of sustaining the analogy must exist in advance somewhere.
>
> (p. 55)

Can this trait, sustaining all analogies, be the copula provided by the verb *is*, the excess of meaning underlined by the categorial intuition, the possibility of emergence through synthesis, the possibility of all possibility? The trait could precisely point to a space, not simply of difference but also of connectedness and of nearness that gives the multiplicity of beings, Being and the ontico-ontological difference. Conceived as such, metaphoricity establishes the movement of nearness that has always already begun, allowing the coming together of beings and the emergence of new beings. This, in consequence, means that Being; namely, that which allows beings to present themselves as they are, and, in other words, the letting relationship that I have been describing from the beginning, is an imaginative process connecting contexts, detecting analogies and allowing the nearing of domains, even when analogies are not readily detected, precisely because there are no strict boundaries and distinctions between areas of meaning as the metaphysical conceptual system would suggest. This understanding of Being resembles the Kantian description of imagination as synthesis that joins and constructs the inner and the outer realms, space and time, appearance and concept, empirical and pure intuition. The precise way this connectedness takes place is the next step of our investigation: indeed, if 'language is the house of being', we need to think the analogies emerging between phrase (a) and phrase (b), arguing that:

a   The human being *is* and dwells in a house that *is*.
b   Language *is* and dwells in a house that *is* and belongs to Being as the process that allows everything to *be*.

Of what sort is the transfer that allows this analogy? Its nature cannot be expressed in propositional terms, since the source domain (a) is completely included in the target domain (b). In other words, the human being is a languaging being deeply defined by its coming to language, and as such it needs to be thought as the being that dwells in the house and at the same time as the one dwelling as the languaging being. What's more, the target domain (b) is possible only as a result of the source domain's (a) being possible and by virtue of both domains' possibility to exist through an analogy to Being, while Being itself exists in analogy to beings, as the one conditioning and mediating this analogy. In other words, the two domains are vast: neither of them is familiar nor traceable; they contain each other, and their relatedness performs the unknown process that conditions Being as possibility. The immensity of these domains suggests that we cannot completely know the kind of crossings and syntheses that take place. Derrida (2007) imagines this process as two parallel lines, namely, thinking and poetry, coming into each other's proximity and ultimately in their common neighbourhood, commenting that:

> Breached and broached, the two parallels cut each other at infinity, recut and confirm each other, notch each other and each signs in some way in the body of the other, the one in the place of the other. They sign there the contract without contract of their neighbourhood.
>
> (p. 74)

Imagery, metaphors of dwelling and spatial metaphors in general come once more to the rescue and, just like Heidegger's metaphors, they do not constitute propositional logic. Instead, they comprise questions proposing a certain orientation of thinking that does not name; it does not answer or conceptualise, but rather stays near to the processes in question. Heidegger, however, is determined to see nearness taking place through language and indeed poetic image. He has claimed, after all, that this nearness comes mostly in sight with language, and also when language's ability to name a thing breaks down. However, this breaking down is most indicative of the very process that makes any meaning possible. On this point, Heidegger (1982b) remarks in *The Nature of Language*: 'But when does language speak itself as language? Curiously enough, when we cannot find the right word for something that concerns us, carries us away, oppresses us or encourages us' (p. 59).

Taking this into account, we can infer that language unfolds differently from the process that gives language just as technology unfolds differently from the process that gives technology: when a table, for example, breaks down, we catch a glimpse, for the first time, of the process that allows the table to be a table, and this process has nothing to do with tables as finished products and furniture. When a table is broken, we get to see the wood, the nails, the paint and possibly the designing and crafting.[2] Indeed, we might first come to recognise the process of gathering that allowed all the elements to come together and to even imagine the absent elements that contributed to the emergence of the table *as* table. Some of these elements are the carpenter, the tree and its environment, the design process, the time of contemplating and manufacturing. Thus, through the absent word and the broken table, we first get to see what gives language and what gives technology, and in both cases the *giveness* itself is neither language nor technology. What the word and the table reveal and reveal by way of hints and gestures, are the processes of exteriorisation that metaphoricity instantiates. In what follows, this is investigated further.

3. The exteriorised and embodied metaphoricity

Metaphoricity as an essentially exteriorising process entails the space of existence that is never a meaningless sensuous familiar something but an inherently meaningful space – linguistic or technological, allowing for the formation of images and of the forming power of the imagination.[3] Metaphoricity's necessary exteriorisation can be traced back to Nietzsche's (2000) position, as expressed in *On Truth and Lie in An Extra-Moral Sense*:

> The 'thing in itself' (which is precisely what the pure truth, apart from any of its consequences, would be) is likewise something quite incomprehensible to the creator of language and something not in the least worth striving for. This creator only designates the relations of things to men, and for expressing these relations he lays hold of the boldest metaphors. To begin with, a nerve stimulus is transferred into an image: first metaphor. The image, in turn, is imitated in a sound: second metaphor. And each time there is a complete overleaping of one sphere, right into the middle of an entirely new and different one.
>
> (p. 55)

Nietzsche's reconceptualisation of metaphoricity opens up a new path into the investigation of human understanding. This kind of metaphoricity, instead of being decorative, linguistic, and at the margins of cognition, is organic, pre-hermeneutical and pre-phenomenological, designating an experience lived by an embodied being that exists in an environment and in connection to this environment. In Nietzsche, metaphoricity is, at least during its first stage of unfolding, an automated connectedness to the world and a type of responsiveness that is not hermeneutic but is ultimately conditioning *hermeneia*. During the first stage of this connectedness, a certain kind of interiorisation modifies the exterior and takes place through the image. The world is received as image. This process could have easily been described as *representing*, but it lacks the neutrality that representation entails. On the contrary, this image-forming process is a metaphorical process of nearing, allowing *Dasein* to be transferred outside itself and to become part of this outside. Without this constitutive movement, there would not be either a world or a being who speaks of this world. Metaphor brings the human being near to other beings and the human being becomes the nearing being through this nearness.

The specific embodiment of the human being, Nietzsche (2000) comments, affects its perception of the world, asserting that 'the insect or the bird perceives an entirely different world from the one that man does' (p. 58). In other words, the human body is always already a form of schematisation, instantiated in the exterior and in*forming* the interior. Therefore, Nietzsche, contrary to Heidegger, understands language as just one stage of metaphoricity's multifaceted unfolding. Even so, the two philosophers seem to agree on the importance of image for metaphoricity – that is to say, image understood as a dynamic process of synthesis and as the first step towards an abstraction, ultimately, leading to metaphysics, representational thinking and nihilism. Indeed, Gayatri Chakravorty Spivak (1997) and Sara Kofman (1993) purport that Nietzsche's specific understanding of metaphor is later transformed into the 'will to power',[4] namely, the force that assimilates the unknown into the known. The movement of transposition is very important for this notion, and thus Kofman argues:

> The transposition is achieved by carrying over the 'known' on to the unknown. It presupposes an activity of assimilation, of digestion, of reducing differences, which is a fundamentally 'unjust' will to mastery (Nietzsche later calls it 'will to power'). Operating already at the organic level, it is still present in intellectual activity, which is supposedly disinterested and at the highest level: the will of the 'mind', too, is to achieve unity out of diversity, to restrict and subjugate the unfamiliar.
>
> (p. 33)

Nietzsche (2000) transforms metaphoricity into a force that allows the interaction of beings. The human being, Nietzsche seems to argue, plays no special role in the process that ultimately constructs truth. Truth, for Nietzsche, is nothing but

> [a] movable host of metaphors, metonyms, and anthropomorphisms: in short, a sum of human relations which have been poetically and rhetorically

intensified, transferred and embellished, and which, after long usage, seem to people to be fixed, canonical, and binding. Truths are illusions which we have forgotten are illusions; they are metaphors that have been drained of sensuous force, coins which have lost their embossing and are now considered and no longer as coins.

(p. 56)

Nietzsche's conceptualisation of metaphoricity comes very close to Heidegger's metaphysical construal of metaphor or to his metaphoric construction of metaphysics. The fact that metaphoricity is later transformed into the Nietzschean will-to-power, which is then easily converted by Heidegger into modern technology's essence, should not come as a surprise. For Heidegger, technology is, similarly to metaphor, the power that turns the particular into the identical and in turn into the abstract.

Nietzsche's contribution to the reconceptualisation of metaphoricity, however, cannot be simply constrained within the boundaries of this negative evaluation, since the philosopher describes, as already noted, a process that allows the projections and the transformations of domains, taking place through the organic, the phenomenal, the automated and the hermeneutic aspects of being. In turn, this multidimensional metaphoricity allows us to rethink the contribution of organic and non-organic matter to the constitution of meaning[5] and to consider the possibility that certain conditioning of *hermeneia* takes place at the non-hermeneutic level. This kind of metaphoricity is instantiated through the coupling of *Dasein* and its environment, taking place via language, technology and the human body. In *Parmenides* (1942–43), Heidegger (1998) seems to discuss for the first time the possibility of such a synergy, underlining the hand's involvement in every possible human activity. He writes:

> Man himself acts [*handelt*] through the hand [*Hand*]; for the hand is, together with the word, the essential distinction of man. Only a being which, like man, 'has' the word (μῦθος, λόγος), can and must 'have' 'the hand.' Through the hand occur both prayer and murder, greeting and thanks, oath and signal, and also the 'work' of the hand, the 'hand-work,' and the tool. The handshake seals the covenant.
>
> (p. 80)

In *Antigone*, language – and language alone – was considered to be the (im)possible origin of humanity, excluding thus the human body. Similarly, in *Being and Time (BT)*, the hand was obsessively mentioned in connection to the ready-to-hand and to the present-at-hand but was in itself ignored. Now, in *Parmenides*, Heidegger finally unearths the strong connection existing between hand and language, proceeding thus to a deconstructive reading of the 'origin' of the human being. He writes:

> The hand sprang forth only out of the word and together with the word. Man does not 'have' hands, but the hand holds the essence of man, because

the word as the essential realm of the hand is the ground of essence of man. The word as what is inscribed and what appears to the regard is the written word, i.e., script. And the word as script is handwriting.

(p. 80)

In *Parmenides*, Heidegger appears to deconstruct the primacy of the spoken word previously understood as the main material of the house of Being. He also locates the human being's essence in writing, asserting that 'the hand holds the essence of man'. This kind of interpretation is not, however, drawn to its conclusions, since, throughout the later period, Heidegger maintains the primacy of language and seems to argue that it is precisely language that makes the hand, whilst admitting that the hand can give writing through language. For Bernard Stiegler (1998), conversely, the interaction between language, hand, and tool, is indicative of something more fundamental: relying on the paleontological evidence discussed by André Leroi-Gourhan (1993), Stiegler argues that humanisation occurs when the hand grasps the tool and the mouth is freed for language. As follows, and 'with primordial consequence', Leroi-Gourhan (1993) asserts that 'tools for the hand, language for the face, are twin poles of the same apparatus' (cited in Stiegler, 1998, p. 145). Tools and language are ultimately co-responsible for the 'specific cerebral organization' that makes up the human being (Stiegler, 1998, p. 145). Stiegler (1998) then infers that,

[i]f paleontology thus ends up with the statement that the hand frees speech, language becomes indissociable from technicity and prostheticity: it must be thought with them, like them, in them, or from the same origin as theirs: from within their mutual essence.

(ibid.)

Metaphoricity can articulate such a synergy: it is an original exteriorisation of the inside, a spatialisation of the temporal realm, and an interconnectedness of all these domains and processes. Stiegler insists that the human being exteriorises itself through the prosthesis of technology and that this exteriorisation is 'a putting-outside-the-self that is also a putting-out-of-range-of-oneself' (p. 146). It is only through exteriosation that the human being pursues its incomplete nature, that is, through the transferring of the self onto exterior supports, often of its own making. Stiegler argues then that,

[w]ith the advent of exteriorisation, the body of the living individual is no longer only a body: it can only function with its tools. An understanding of the archaic anthropological system will only become possible with the simultaneous examination of the skeleton, the central nervous system, and equipment.

(p. 148)

Heidegger (1998), of course, recognises the important relation that connects the hand with language. However, he also sees modern technological artefacts, like the typewriter, intruding and corrupting this key relationship by turning that which is meaningful into meaningless object, calculative distance and distance-lessness. He thus writes:

> The typewriter tears writing from the essential realm of the hand, i.e., the realm of the word. The word itself turns into something 'typed.' Where typewriting, on the contrary, is only a transcription and serves to preserve the writing, or turns into print something already written, there it has proper though limited, significance. [. . .] Mechanical writing deprives the hand of its rank in the realm of the written word and degrades the word to a means of communication. In addition, mechanical writing provides this "advantage," that it conceals the handwriting and thereby the character. The typewriter makes everyone look the same.
>
> (p. 81)

Heidegger's analysis, assuming here intense Marxist overtones, suggests that modern technology brings about estranged labour and alienation. The image, as the something typed, is instrumental in this process: similarly to his previous discussions of the modern *polis*, Heidegger seems to claim once again that the revealing, specific to modern technology, does not bring about nearness. A writing implement, like the typewriter, turns the word into image; it sacrifices individuality for the sake of efficiency and brings about a specific exteriorisation of the self that postpones the potential of imagination. Before addressing this point head on, let me, bring to the fore my critique concerning Heidegger's analysis and its evident lack of concern for the learning process that precisely allows the ready-to-hand to become ready-to-hand. This eminent absence, to which I pointed earlier, is finally, addressed in *What is Called Thinking?* In this text, Heidegger (1968) describes the learning process experienced by the cabinetmaker's apprentice, asserting that:

> His learning is not mere practice, to gain facility in the use of tools. Nor does he merely gain knowledge about the forms of things he is to build. If he is to become a true cabinetmaker, he makes himself answer and respond to the different kinds of wood and to the shapes slumbering within wood to wood as it enters into man's dwelling with all the hidden riches of its nature. In fact, this relatedness to wood is what maintains the whole craft. Without that relatedness, the craft will never be anything but empty busywork, any occupation with it will be determined exclusively by business concerns. Every handicraft, all human dealings are constantly in that danger. The writing of poetry is no more exempt from it than is the business of thinking.
>
> (pp. 14–15)

He then adds that:

> Every motion of the hand in every one of its works carries itself through the element of thinking, every bearing of the hand bears itself in that element.

All the work of the hand is rooted in thinking. Therefore, thinking itself is man's simplest, and for that reason hardest, handiwork, if it would be accomplished at its proper time.

(pp. 16–17)

Heidegger's desire to differentiate between the calculative thinking that epitomises the essence of modern technology and meditative/poetic thinking or *Gelassenheit*, draws him closer to some type of a phenomenological explication of the tool's familiarity that also brings to the fore metaphoricity's function as embodied and material exteriorisation. In the *Zollikon Seminars*, Heidegger (2001) expands on the topic of embodied exteriorisation, asserting that:

When I grasp the glass, I not only grasp the glass, but can also simultaneously see my hand and the glass. But I cannot see my eye and my seeing, and by no means am I able to grasp them. For in the immediacy of seeing and hearing turned toward the 'world,' the eye and ear disappear in a peculiar manner. If someone else wants to ascertain how the eye is functioning when seeing, and how it is anatomically constituted, he must see my eye as I see the crossbar.

(p. 82)

The phenomenon of 'double sensation [*Doppelempfindung*]' expresses best the exteriorisation of the self that allows the perception of 'what is touched and the sensation of my hand' – as the one doing the touching and as the one being touched (p. 83). This phenomenon differentiates the hand from the eye since, '[i]n grasping, the hand is in immediate contact with what is grasped', whereas '[m]y eye is not in immediate contact with what is seen' (pp. 109, 183). The eye grasps the hand grasping, and, in this grasping, the eye grasps a notion of a self. The phenomenon as such suggests that exteriorisation – that is to say, the very process of becoming human – necessitates the hand so as to instantiate intention by being-(the)-outside-(of)-the-human. This exteriorisation reveals the interiority that initiated the action but is itself constituted and formed by this very process. The hand therefore creates immediacy through the experience of distance. In contrast to the eye's functioning, the hand grasps the glass, only because it is able to experience distance; the hand *brings-near by preserving distance*. This process takes place with objects such as glasses, nails and computers, but it also takes place with my own hand, since when I grasp my hand grasping a thing, I turn this thing into something graspable. But, does this process constitute the original meaning of *Dasein*? Is *Dasein* not the very being that constitutes its own *there* and whose *there* is always already part of the world? In fact, is this mode of thinking not the only way that *Dasein* can be conceptualised as subject? *Dasein* is the being that grasps the world and interprets it in the very moment when it sees itself being grasped and being interpreted by the world. Is this functionality of the hand not a basic form of interpretation? Is the hand not the very thing that lets things be (*lassen*) by letting them come near? Is the hand, with its unique ability to experience nearness in preserving distance, not the human being's first differential possibility – that is, the first affirmation of distance between an I and its world that makes the I a possibility of differentiation from this world, which is not me but already part of me? Finally, is the hand not the very first possibility of transference out of a self that is I but always already in the process of becoming a different I? If the above questions are answered positively, then a different question arises: Is there a developmental process of learning that is similar to the transfer (*metaphora*) of the self, namely, a metaphor that has in any case always already happened, being itself a movement beyond this self through the becoming of this self and another self, namely, through the becoming of glass, nail, computer and ultimately the world? Is this bringing-near an important *metaphor*, a transfer of the non-yet-self onto the things it brings near?

In consideration of this possibility, the *metaphor* of the *house of Being* moves beyond the distinctions between theory and practice, sensible and intelligible, and brings together language and technology. Both language and technology are implicated in the process of *differential metaphorical learning*. When, for example, an infant grasps her blanket she does not know the word for grasping and as such the action of grasping cannot be present *as* the meaning of this word. The action, however, is always already metaphorical – the infant is transferred onto things – long before metaphor is understood literally as a literary term. The action is also metaphorical to the extent that there is no previous literal meaning of this action and as such it occupies a middle ground in the constitution of meaning. The body contributes to the constitution of the analogy that lies at the heart of metaphor; every time the child chooses to use one hand or the other, to move back or forth, to jump up or down, they do so in analogy to a possibility. They enact différance (Kouppanou & Stiegler, 2016). Similarly, the experience of looking at the mirror – that is, being transferred into the image and being that image in analogy to being-here – ultimately constructs the feeling of being-here-there and constitutes the unfolding of difference through metaphor; it is being in différance. Every time the young child chooses to say the word for the blanket – when, that is, they are able to say it – or each time they choose to grasp the blanket, they participate in différance; the child can either say it *or* grasp it. In consideration of this process, technology becomes quite important: it allows the complicated interaction of embodiment and language. This type of metaphoricity can be very slow and *passive* during which a specific social milieu is constructed through the preservation of time and through its own schematic and metaphoric form that is subsequently learned by individuals, allowing thus the emergence of new ways of living and thinking. In Stiegler (1998), as we have seen, the equivalent of this process of learning is 'epiphylogenesis', and it refers to a recapitulation of past memories. Metaphor, however, can underline the movement of nearing and passive learning that produces new meaning. In this respect, it resembles psychoanalytic *transference*, pointing to the redirection of feelings, the reproduction of thoughts, the imaginative connection between past and present, presence and absence. Furthermore, metaphor, as a process of learning and connectedness to the world, pays heed to the role of image, which is not simply understood as memory, and to the role of the tool that instantiates schematic and metaphoric powers. In what comes next, a new conceptualisation of the tool is presented; indeed, one which is in accord with Heidegger's implicit use of metaphor and with his explicit discussion of nearness.

4. The metaphoric power of tools and nearness

In the essays, *The Thing* (1950) and *Building Dwelling Thinking* (1951), Heidegger (1975b) addresses *nearness* as 'Being itself' (Richardson, 2003, p. 567). Nearness is experienced through activities such as using a jug or crossing a bridge and is juxtaposed to *distancelessness* produced by modern technology and experienced in activities like watching the news. The introductory paragraph of *The Thing* makes a good starting point for this discussion. In this text, Heidegger (1975c) claims that:

> All distances in time and space are shrinking. Man now reaches overnight, by plane, places which formerly took weeks and months of travel. He now receives instant information, by radio, of events which he formerly learned about only years later, if at all. [. . .] The peak of this abolition of every possibility of remoteness is reached by television, which will soon pervade and dominate the whole machinery of communication.
>
> (p. 165)

Heidegger describes the elimination of spatial and temporal qualities of places and the ensuing feeling of distancelessness – that is, the possibility of engagement with multiple locations and things that turn meaningful spacetime into calculable undifferentiated distance. The shortening of distance, nearness, and remoteness condition our existential and thinking modalities, which are in any case technologically mediated. However, in Heidegger's later thought, it is believed that modern technology's mediation replaces one set of experiences with another: knowledge with information, natural phenomena with media representations, otherness with TV programmes, and originary image with assimilation. This displacement creates the illusion of the Other's accessibility and of nearness, but the truth is that this newfound otherness is quite different from the one discovered through the originary image. It is rather incapable of making the world unfamiliar.

With new technologies, nearness, as that which allows things to reveal themselves, withdraws, while beings fall into a 'uniformity in which everything is neither far nor near' (Heidegger, 1975c, p. 166). Uniformity refers to phenomena like the leveling of connectedness we come to experience when watching, for example, a report on a natural disaster taking place far away from us. If we do not have any previous connectedness to the area or a prior sensitivity to this matter, the event will probably stay far away from us, even though it will be experienced as something close and familiar. This closeness, however, does not involve us or alter our concerns; it does not address us, and it certainly does not transform us. This kind of numbness has, however, a further effect, since it can potentially transform all experiences into uniform distancelessness. This is, in fact, the effect that Heidegger considers as modern technology's greatest threat: technology threatens us with the dominion of a single frame of mind that suspends our capacity to receive things *as* the things that they are and instead enforces previous perceptions on them.

Modern media was, however, an area of being that Heidegger's analytic (2008) addressed as early as *BT*, stating there that '[w]ith the 'radio' [. . .] *Dasein* has so expanded its everyday environment that it has accomplished a de-severance of the "world" – a de-severance which, in its meaning for *Dasein*, cannot yet be visualized' (p. 140/105). Having in mind, that this earlier analysis constitutes a peculiar middle ground resting between a negative evaluation of technology and an ontological exegesis of eternal structures of existence, we cannot be certain of Heidegger's intentions. Does this early postponement suggest a hesitation concerning the possible negative nature of all tools? The ready-to-hand is, after all, understood on the basis of its potentiality to extend *Dasein*'s temporal reach, and this is precisely what modern media are purported to do best. Heidegger's example of *Dasein*'s discovering of other *Daseins* in the process of creating shoes is indicative of this point. *Dasein* discovers others as it uses tools. In the case of modern media, however, the discovery of others is not a mere by-product of technology but precisely the very product in demand. *Dasein* watches the news and listens to the radio in order to know about other *Daseins*. For Heidegger, this knowing constitutes nothing but curiosity. Now, if we transfer this critique to the sphere of modern social media, we can easily infer that the curiosity intensifies, and the desire for discovering and for connecting with others is transfigured into an overwhelming desire for nearness. It will not be satisfactory, however, if I limit here my analysis of social media to a reiteration of Heidegger's critique of modern technology, since such an analysis would not have taken into consideration the seminal notions of imagination, metaphoricity and schematic power. Before I proceed to such an analysis in the next chapter, however, we need to discuss how later Heidegger changes the way he thinks about tools.

Later Heidegger attempts to think nearness through things that once participated, or perhaps still do, in experiences of true nearness. For Michael Lewis (2005) this possibility is realised through Heidegger's phenomenology of things. As he puts it:

> Today, this unifying trait of 'beingness' is named by Heidegger as 'techniciz-ability' or 'makeability', and it is precisely those singular and fragile beings called 'things' which technology is by definition unable to make, this inability marking its own blind spot.
>
> (p. 81)

Through this turn to phenomenology, Heidegger mostly discusses crafts, that is, technologies of a different era, or, as Lewis puts it, things that are not made by technology, and this in consequence raises the following questions: Is it only pre-industrial things that afford true nearness? Is nearness afforded by our own frame of mind? Does a change of mentality – our performing of *Gelassenheit* – modify the way we relate to things? And if so, can a modern object like a social networking site afford true nearness and be an originary image of some sort? Is there a possibility for digital objects to be used without imposing the very mathematical mentality that produced them? And, can their own mode of connectedness coexist with or even be a type of poetic thinking?

Heidegger (1975c) shows the path toward the contemplation of these topics in *The Thing* in which text Heidegger looks into the nature of nearness 'by attending to what is near', that is, to common things like a jug (p. 166). Heidegger understands these things as gatherings while mentioning that the jug's presence comes to light with its use, namely, the outpouring that Heidegger calls 'the poured gift' (ibid.). In this way, the analysis diverges from the perspective followed in *BT*. Instead of considering the tool as a mere node connecting with other nodes in a network of usable tools, people, and projects, it is now believed that the tool instantiates a meaningful focal node gathering a single time-space. This is a process of nearing, but contrary to the earlier emphasis on the *towards-which* and on the future, the emphasis is now placed on the tool's gathering function. This new conceptualisation of the tool puts into question Heidegger's earlier analysis of technology and especially his conceptualisation of the home, or lack thereof. Indeed, a phenomenology that conceives the home as gathering and it is itself devoid of any ideological content, is precisely what is missing from *BT*.

In the gift of the jug's outpouring, Heidegger says, we can observe the respectful receptiveness of things and the gathering of the elements of the *fourfold*. The fourfold is a very poetic and 'mythic' notion in later Heidegger, but its 'philosophical import [. . .] is far from clear' (Olafson, 1993, p. 117). For this reason, there are those who dismiss it 'as an example of pious gibberish' and those 'who leave it out of consideration altogether' (Harman, 2002, p. 190). There is, also, a third group of scholars who support that the strength of the fourfold lies in its power to offer a kind of preparation for an alternative, 'deconstructive living', since '[t]alk of the Fourfold has the advantage of bringing in neglected dimensions of the event of the world, and it avoids the metaphysical temptation to interpret man, things and being in terms of constant presence' (Kolb, 1986, pp. 184,191).

At this point, however, we need to pay attention to the way Heidegger discusses the simple action of the outpouring. Indeed, Heidegger (1975c) says that in the outpouring the elements of the fourfold – earth, sky, divinities and mortals – stay true to their nature as they receive one another; earth and sky meet in the water, earth and rain dwell in the rock, the water satisfies the mortals' thirst and also becomes the mortals' sacrifice to the gods. This kind of interaction allows a thing to maintain its own nature: in this gift, all the element of the fourfold gather in their separateness and oneness. As such, this process is quite different from modern technology's tendency to transform every being into resource, energy, and information, locked, stored and consumed. Instead, the gift of the outpouring is transformed into a type of nearness that gathers the fourfold, allowing the interaction of multiplicity through a process called 'mirror-play' (Heidegger, 1977a, p. 179).

As I have already mentioned, there are many interpretations of the four elements, which we cannot possibly present here. Julian Young (2006), for example, argues that earth refers to the 'totality of things, animal, vegetable and mineral, with which we share our world', while sky represents 'the representation of time changes, seasons and the weather' (p. 374). Gods refer to the fundamental *ethos* of a community', while *Dasein*'s mortality describes the 'defining feature of human beings' (p. 375). In a similar fashion, William Richardson (2003) asserts that earth and sky refer to the physical reality but diversely claims that gods invoke 'the entire domain of the divine' (p. 572). He also underlines the mortals' capacity to experience *death as death*, believing that this structure ultimately alludes to the experience of the Being's withdrawal. If this is true, though, we can entertain the possibility that, during this period, Heidegger moves away from subjectivism and voluntarism but maintains an unaltered belief in *Dasein's* essential propensity to interpret: mortals do not perish; they encounter death *as* death and receive the messages from the gods through the as-structure. The as-structure takes us back to Being, temporality and imagination. Now, however, it is co-constituted by the interaction of a multiplicity of elements, the proximities they contrive and the gatherings that the fourfold allows to manifest 'into a single time-space, a single stay' (Heidegger, 1957c, p. 174). In this light, the fourfold appears to be a metaphor of being and a metaphor of thinking, that is, of a type of thinking focused on a single concern. This singularity does not diminish multiplicity. It remains responsive to the things constituting a single concern without succumbing to preconceptions and representations. Therefore, Heidegger writes:

> The thing things. In thinging, it stays earth and sky, divinities and mortals. Staying, the thing brings the four, in their remoteness, near to one another. This bringing-near is nearing. Nearing is the presencing of nearness. Nearness brings near – draws nigh to one another – the far and, indeed, *as* the far. Nearness preserves farness.
>
> (pp. 177–178)

The distinction between nearness and distancelessness becomes very clear. Nearness does not obliterate farness, and it is not farness's opposite. In fact, that which is respected in its remoteness is the very thing that comes near. Nearness is precisely this. Heidegger explains: 'Nearness is at work in bringing near, as the thinging of the thing' (p. 178). This is the event of the world, the 'worlding' of the world (1975a, p. 45). Thinging and worlding are responsive and transformative and can be thus thought as schematic processes that orchestrate our modes of being. Still, one might wonder, especially in the light of the obsessive tendency for connectedness that modern technologies exhibit: Can we make or build things that gather the fourfold? Can computerised and digital objects, such as social networking sites and search engines, gather the fourfold, bring near things and preserve the farness? And if not, how can we think the schematic power characteristic of these tools? I believe that the answers to these questions can be explored through Heidegger's famous essay *Building Dwelling Thinking* (1951), which is discussed in the next chapter.

5. Conclusion

With Heidegger's later phenomenology, the philosopher resumes his investigations into the categorial intuition, the hermeneutical relation, the *letting* relationship, imagination and nearness. All these processes were previously understood through a temporocentric perspective focused on the future. Heidegger now, more than ever, privileges language, which he sees as the only possible way to challenge the everyday inauthentic and technological familiarity of the world, but he also opens up the space for the reconsideration of our relation to technology. His own text reveals the power of metaphor, even though he certainly does not name it as such. His later accounts of mundane things as the jug reveal a new possibility of thinking the tool; indeed, of perceiving it as a modality of schematisation and metaphoricity. Tools work on our thoughts. Technologies make up our minds or, to speak more 'scientifically', build our cerebral organisation and organise our embodiment. The discussion in *The Thing* suggests exactly that – that technology participates in imagination, offering the already-there in which dwelling takes place. In this respect, both the possibility of *alētheia* and the possibility of representation are conditioned by the metaphorical process, which cannot be simply determined by language's or technology's separate involvement.

However, the question still remains: Are there tools inhibiting imagination from taking place? In other words, if we move from familiarity to familiarity, can imagination shut down? Can dwelling be turned into homelessness? To put this in Stieglerian terms: Modern technology provides secondary retentions that form our selection criteria, and this, in turn, defines and conditions that which can be thought by our thought and can be imagined by our imagination. To put it in Heideggerian terms: Modern technologies produce representations that include the subject within what is represented; the user becomes an image, whilst the image conceals its constitution. Lastly, to put this in this book's terms: Machines have specific metaphoric powers that organise their users. However, can this function suspend imagination, différance, and nearness? In the next chapter, I examine this possibility through a focus on digital technologies.

Notes

1 Ziarek (1994) comments that according to Derrida, '[t]he phrase "*das Haus des Seins*" works, then non-literally and non-metaphorically; it is instead a quasi-metaphor, a transfer not *between* words but *into* words, which characterizes language on its way to expression. Derrida concludes that this Heideggerian saying opens in its design, in its incision into Being (*entame, Aufriss*), the space for the conceptual network. Consequently, such quasi-metaphoricity performs the linguistic transfer into the difference between the literal and the metaphorical, where the transfer into the difference between the literal and the metaphorical, where the transfer, the *phora*

associated with metaphor, becomes possible as a result of a quasi-transfer, itself already in retreat (*retrait*), tracing and retracing, marking a mesh of traits (retrait, *Gezüge*)' (p. 18).

2 Heidegger gives such an account of production; indeed, one according to the four – that is, according to materiality of materiality (*causa materialis*), form (*causa formalis*), purpose (*causa finalis*) and the producer (*causa efficiens*), in his essay *The Question Concerning Technology* (1977). There, he asserts that the four causes belong together and are co-responsible for the emergence of things. This account bears similar connotations with his account of the fourfold, in the respect that it describes the emergence of beings in terms of gathering processes that allow the nearing and the responsiveness of different domains.

3 From here on, when I mention imagination, I will be referring to lived time, to nearness and the letting-relationship.

4 I believe that Heidegger's acceptance of the will-to-power as part of the German people's destiny, during the middle period of his thought, says more about his politics and less about his beliefs concerning metaphor. Indeed, the negative consideration of metaphor as the one responsible for the assimilation of meaning is present in the work of both Nietzsche and Heidegger.

5 The role of non-organic matter, and its organisation as technological artifact, has been discussed by Stiegler (1998).

References

Arendt, H. (1998). *The Human Condition* (2nd ed.). Chicago: University of Chicago.

Derrida, J. (2007). The Retrait of Metaphor. In P. Kamuf & E. Rottenberg (Eds.), *Psyche: Inventions of the Other* (Vol. 1, pp. 48–80). Stanford, CA: Stanford University Press.

Dreyfus, H. (1989). Beyond Hermeneutics: Interpretation in Later Heidegger and Recent Foucault. In G. Shapiro & A. Sica (Eds.), *Hermeneutics: Questions and Prospects* (pp. 66–83). Amherst: University of Massachusetts Press.

Gasché, R. (1997). *The Tain of the Metaphor: Derrida and the Philosophy of Reflection.* Cambridge, Massachusetts and London: Harvard University Press.

Harman, G. (2002). *Tool-Being: Heidegger and the Metaphysics of Objects.* Chicago and La Salle, IL: Open Court.

Heidegger, M. (1968). *What Is Called Thinking?* (F. D. Wieck & J. G. Gray, Trans.). New York: Harper and Row.

Heidegger, M. (1969). *Discourse on Thinking* (J. M. Anderson & E. Hans Freund, Trans.). New York: Harper Perennial. (Memorial Address).

Heidegger, M. (1975a). The Origin of the Work of Art (A. Hofstadter, Trans.). In *Poetry, Language, Thought.* New York: Harper and Row.

Heidegger, M. (1975b). *Poetry, Language, Thought.* In A. Hofstadter (Ed.). New York: Harper and Row.

Heidegger, M. (1975c). The Thing (A. Hofstadter, Trans.). In A. Hofstadter (Ed.), *Poetry, Language, Thought* (pp. 165–186). New York: Harper and Row.

Heidegger, M. (1977). The Question Concerning Technology (W. Lovitt, Trans.). In *The Question Concerning Technology, and Other Essays.* New York and London: Harper and Row.

Heidegger, M. (1982a). A Dialogue on Language. *On the Way to Language* (P. D. Hertz, Trans.) (pp. 1–56). Oxford: Harper One.

Heidegger, M. (1982b). The Nature of Language. *On the Way to Language* (P. D. Hertz, Trans.) (pp. 57–110). Oxford: Harper One.

Heidegger, M. (1998). *Parmenides* (A. Schuwer & R. Rojcewicz, Trans.). Bloomington and Indianapolis, IN: Indiana University Press.

Heidegger, M. (2001). *Zollikon Seminars: Protocols- Conversations-Letters* (R. Askay &
F. Mayr, Trans.). In M. Boss (Ed.). Evanston, IL: Northwestern University Press.

Heidegger, M. (2008). *Being and Time* (J. Macquarrie & E. Robinson, Trans.).
Oxford: Blackwell.

Heidegger, M. (2010). Letter on Humanism. In D. F. Krell (Ed.), *Basic Writings*
(pp. 147–181). London and New York: Routledge Classics.

Kofman, S. (1993). *Nietzsche and Metaphor* (D. Large, Trans.). Stanford, CA: Stanford University Press.

Kolb, D. (1986). *The Critique of Pure Modernity: Hegel, Heidegger, and After*. Chicago and London: Chicago University Press.

Kouppanou, A., & Stiegler, B. (2016). ". . . Einstein's Most Rational Dimension
of Noetic Life and the Teddy Bear . . ." An Interview with Bernard Stiegler on
Childhood, Education and the Digital. *Studies in Philosophy and Education, 35*(3),
241–249.

Lewis, M. (2005). *Heidegger and the Place of Ethics*. London: Continuum.

Nietzsche, F. (2000). On Truth and Lie in an Extra-Moral Sense. In C. Cazeaux
(Ed.), *The Continental Aesthetics Reader* (pp, 53–62). London and New York: Taylor and Francis.

Olafson, F. (1993). The Unity in Heidegger's Thought. In C. B. Guignon (Ed.),
The Cambridge Companion to Heidegger (pp. 97–121). Cambridge: Cambridge
University Press.

Palmer, R. (1984). On the Trancendability of Hermeneutics. In G. Shapiro & A. Sica
(Eds.), *Hermeneutics: Questions and Prospects*. Amherst, MA: University of Massachusetts Press.

Richardson, W. J. (2003). *Heidegger: Through Phenomenology to Thought*. New York:
Fordham University Press.

Spivak, G. C. (1997). *Translator's Preface of Grammatology*. Baltimore and London:
The John Hopkins University Press.

Stellardi, G. (2000). *Heidegger and Derrida on Philosophy and Metaphor: Imperfect
Thought*. New York: Humanity Books.

Stiegler, B. (1998). *Technics and Time, 1: The Fault of Epimetheus* (R. Beardsworth &
G. Collins, Trans.). Stanford, CA: Stanford University Press.

Young, J. (2006). The Fourfold. In C. B. Guignon (Ed.), *The Cambridge Companion
to Heidegger* (pp. 373–392). Cambridge: Cambridge University Press.

Ziarek, K. (1994). *Inlfected Language: Towards a Hermeneutics of Nearness: Heidegger, Levinas, Stevens, Celan*. Albany, NY: SUNY Press.

7 Digital metaphoric machines of nearness

1. Introduction

An originary image is an image that presences originarily. Such conceptualisation of image parts with the notions of presence, originary substance or essence, and directs our attention to modes of presencing. What's more, with this conceptualisation, the binary between languaging/poetic image and technological/representative image can only be sustained if language and technology are substantiated as stable modes of presencing, not able to participate, not by any means, in differentiated modes of revealing. This book's phenomenological perspective supports the opposite view, entertaining the possibility that languaging and technological implements can potentially exhibit and participate in differentiated modes of presencing. This deconstructive reading was undertaken here, not in order to replace Heideggerian critique with utter relativism, but in order to uncover underlying processes influencing both technology and language. The very criticism that I voiced against Heidegger's later theory of technology dealt precisely with this point: that our inability to see differences – a perspective that the discourse of Enframing presupposes – does not liberate us but condemns us to a 'seeing' that sees no differences even in cases that such differences are attempting to articulate themselves. The very ground of the articulation of difference, or Derrida's différance, was understood here as permeated with imaginative and metaphoric synthesis constituting what Heidegger calls nearness. Indeed, I have argued that Heidegger's investigations of the Kantian imagination led to the discovery of the workings of time-synthesis in terms of a process of schematisation that functions on the premise of bringing distinct domains close to each other. This is how the present moment – the now – comes to be: through the nearing of the past towards a possible future. Past memories are projected onto and weaved into the expectation of what is to come and therefore transform it by giving us our present. Time comes to *be* by means of this nearing.

Nearness is important for Heidegger – pursued steadily and piously. However, by attempting to define a predominantly spatiotemporal process in terms of time and time alone, Heidegger managed to show something essential concerning the workings of time; namely, its metaphorical nature: time is experienced and understood as nearness, precisely because it is transformational. At the beginning of Heidegger's thought, this nearness is one-dimensionally thought as the privileged pole of each and every possible metaphysical dichotomy: it is understood as temporal rather than spatial, as linguistic rather than technological, and as future-oriented rather than present-oriented. At the heart of all these dichotomies, lies the belief that authentic nearness is defined by the finality of death rather than the technologically mediated finality of daily projects. This reading, in consequence, has turned nearness into an one-dimensional temporality: We are always already preoccupied with one single concern. We are always already unfolding as beings in this process defined by the fact that at any time we cannot be close to more than one thing – this was, after all, Heidegger's explanation concerning the orientation of the human being; namely, that we cannot be simultaneously preoccupied with that which is on our right hand and with that which is on our left hand. Nearness is always wanting in early Heidegger.

In middle Heidegger, nearness becomes, similarly to homeland and to *polis*, a centripetal force and an impossible limit that disregards exteriority, technology and dispersion. This exclusion of otherness is ontological, epistemological and deeply political, leading Heidegger to eliminate imagination from his conceptualisation of time. In contrast, in later Heidegger's phenomenology of things, nearness is articulated in terms of polyphony, movement and inclusiveness. Heidegger's description of the fourfold constitutes in fact an attempt to investigate an ontology of nearness that highlights the inherent dispersion of things. Such a movement is metaphorical. It is thought as a process that allows the approximation and the transformation of domains, which are not in any case pure concepts, but rather complicated messy networks, experiences, and stories that mirror each other selectively and creatively, leading to the emergence of ever new processes and realities. In summary, Heidegger sketches a concept of nearness that functions as metaphoricity, even though it is never named as such. Through the specific perspective taken in this book, metaphoricity is understood, throughout the various stages of Heidegger's thought, as:

a   *Time-synthesis.* Time is a metaphorical process in the respect that past memories are *projected* onto the future, forming anticipation and transforming the now. The now is a continual process of becoming, never something complete, always something in the process of being formed (*Phenomenological interpretation of Kant's Critique of Pure Reason, Being and Time*).

b   *Aesthesis.* A metaphor unfolding as aesthesis, connecting a material thing with a human sense organ. Metaphor is perception (*The Principle of Reason*).

c   Unfolding between a human body and a material thing as in the case of a hand grasping a pair of glasses. The body is transferred onto the tools it uses (*Zollikon Seminars*).

d   Unfolding between a human hand, a writing tool, and language, that is, as the exteriorisation of the self through technology and language (*Parmenides*).

e   Unfolding between different elements and between elements and the human being, as is the case with the fourfold (*The Thing*).

f Unfolding between different unbounded domains containing each other, as with the case of the phrase, 'language is the house of Being' (*Letter on Humanism*), in which we can imagine a human being dwelling in a house and entering a mirror-play with a language-being dwelling in the house of Being while all four elements constitute an 'oneness'.

g Constituting metaphysics through the construction of new conceptual domains (*The Principle of Reason*).

h Allowing the abandonment into the nameless: We do not substitute one word with another or a name with a different name even though at some level and in some cases this happens as well. This conseptualisation of metaphor resists propositional logic by holding together two or more activated and distinct domains, experiences, or networks of meaning. In I. A. Richards, (1936, cited in Ricœur, 2004) this process takes place through the tenor to which we attribute aspects and characteristics of the vehicle. In the metaphor 'Juliet is the sun', for example, some attributes of the sun, like, brightness, warmness, liveliness and so on, are *selected* among the characteristics of the schema of the literal sun and are *projected* onto Juliet. In a similar fashion, Max Black's (1962) interactionist theory of metaphor affirms that there are two parts in a metaphorical sentence: the focus, namely, the word used metaphorically, and the frame, that is, the rest of the sentence. For Black, it is the interaction between the focus and the frame that brings forth new meaning and not the mere substitution of a literal word with a metaphorical one. In other words, the frame *filters* the focus, allowing some of its aspects to shine through and forcing others to remain in the dark. Similarly, cognitive linguists, Lakoff and Johnson (1980), argue that: 'The very systematicity that allows us to comprehend one aspect of a concept in terms of another (e.g., comprehending an aspect of arguing in terms of battle) will necessarily hide other aspects of the concept' (p. 10). The *blending theory* of cognition presents this process in a similar light, while its inaugurators, Gilles Fauconnier and Mark Turner (2002), argue that, with conceptual blending, two distinct mental spaces, which do not refer to rigid conceptual domains, but rather to messy schemas, interact *selectively* allowing the *emergence* of a new domain called *blended space*. Fauconnier (2001) explains it in the following terms:

> A nice example of conceptual blending in action and design is the 'desktop' interface, in which the computer user moves icons around on a simulated desktop [. . .]. Users recruit from their knowledge of office work, interpersonal commands, pointing, and choosing from lists. All of these are 'inputs' to the imaginative invention of a blended scenario that serves as the basis for integrated performance. [. . .] The blend is not the screen: the blend is an imaginative mental creation that lets us use the computer hardware and software effectively.

Fauconnier's belief in the immaterial nature of the conceptual blend clearly indicates that much of the work produced within these innovative theoretical frameworks maintains the belief that the medium's material schematising power is irrelevant. In contrast, I believe that the screen is not 'an imaginative mental creation' and that without its specific material instantiation such a blend would have not been able to exist in its given form, forming power, and functionality (Kouppanou, 2016). What's more, we cannot think of any designing imagination not taking into account the precise instantiation of the screen and the specific nature of the human embodied user as the hands-body-positioning, eyes-looking, fingers-keyboarding being. In other words, I argue that materiality instantiates bodily schemas and this affects the ways we think about technologies and the ways we think about thinking in general. Katherine N. Hayles (2002) is amongst the few scholars addressing the effects of materiality on thinking and not surprisingly names a thing like a book, a 'material metaphor, a term that foregrounds the traffic between words and physical artefacts' (p. 23).

Despite the apparent marginalisation of materiality in Fauconnier and Turner (2002), I find their theory extremely insightful, especially, in the respect, that it does not presuppose strictly defined conceptual representations or pure concepts, but rather complicated schemas or scenarios. Following this tradition partly established by him, Mark Turner (1996) insists on the messiness of domains and proposes that the human understanding should be conceptualised in terms of projection; indeed, in terms of stories projected onto other stories, arguing that our minds create meaning by projecting an 'overt source story onto a covert target story' (p. 6). His perspective therefore abandons the notion of the concept and reorients itself towards the story, the poem, and the allegory. He says:

> Parable begins with narrative imagining – the understanding of a complex of objects, events, and actors as organized by our knowledge of *story*. It then combines story with projection: one story is projected onto another. The essence of parable is its intricate combining of two of our basic forms of knowledge – story and projection. This classic combination produces one of our keenest mental processes for constructing meaning. The evolution of the genre of parable is thus neither accidental nor exclusively literary: it follows inevitably from the nature of our conceptual systems. The motivations for parable are as strong as the motivations for color vision or sentence structure or the ability to hit a distant object with a stone.
>
> (p. 5)

With Turner's emphasis on *projection*, we come full circle, back to Heidegger's rereading of Kantian imagination, synthesis, and time; indeed, back at the beginning of our investigation into the nature of originary image and *alētheia* and back into Plato's *Allegory of the Cave*. There, *alētheia* was discussed through metaphor, through allegory, and through the projection of one story – the story of the cave – onto another story – the story of truth and education. What is most impressive, however, is that this foundational Platonic text, concerned with the indispensability and the eternality of concepts, relies on metaphor and allegory in order to sketch the concept of truth. Is truth then nothing but stories? And, is Heidegger, namely, the one attempting to move beyond Plato and beyond (*meta*) metaphysics, not doing precisely the same thing, namely, telling stories? Indeed, what else, but an image and a story, is his metaphor about the house of Being? Is it not the projection of a story – about human beings living in a house and in language – onto a story about language itself living in a house, told by the very being that needs to think Being? (*Letter on Humanism*).

In summary, Heidegger's most important metaphor, and most important discussion about metaphor for that matter, describes perfectly the notion of nearness: This process is neither language nor technology; it is rather that which gives language and technology or perhaps that which participates in language and technology or even allows for their interaction. Such a process is essentially circular and hermeneutical: the human being, namely, the languaging being, utters a language about language while attempting to understand itself in and as language. In the same vein, the human being, the dwelling being par excellence, thinks about dwelling, while dwelling conditions this very thinking. Heidegger attempted to understand these interactions through a lens, different from calculation and mathematical thinking – namely, away from notions that he attributed to technology. These interactions, however, are possible through the very technologies of being, both dwelling and language. Heidegger's insistence on the notion of aesthesis, indicates the interconnection between the material, and specifically instantiated figuration of the symbolic, the mythological and the imaginative realms, with the narrative renditions of being. This interaction binds language to technology as processes that interact with each other but as also defined, separately and together, by the same underlying process; namely, figuration. In what comes next, I offer my final points on figuration and metaphoricity.

2. The power of figuration

Bernard Stiegler relies on paleontological evidence, offered by André Leroi-Gourhan (1993), in order to ground humanisation and différance on the word/tool synergy, asserting that '[l]anguage is an artefact' as well (Kouppanou & Stiegler, 2016, p. 245). Similarly, I turn now to evidence offered by Leroi-Gourhan in order to uncover the importance of figuration for humanisation. More specifically Leroi-Gourhan (1993) understands figuration as

> part of the same human aptitude, that of reflecting reality in verbal or gestural symbols or in material form as figures: Just as the emergence of language is connected with that of hand tools, figurative representation cannot be separated from the common source from which all making and all representation spring.
>
> (p. 363)

Following Leroi-Gourhan's perspective, we can infer that figuration is intertwined with exteriosation: it connects gesture with word, word with tool, tool with script, constituting the synergetic process that allows for the emergence of meaning. My main argument here underlined precisely this point: that transformative metaphoricity takes place at the embodied, linguistic, technological, and conceptual level, necessitating the interaction of some or of all of these aspects of existence. Leroi-Gourhan (1993) fortifies this position, arguing that the human capacity for abstract thinking arises from figurative behavior manifested as art, language and writing, explaining that

> in its origins figurative art was directly linked with language and was much closer to writing (in the broadest sense) than to what we understand by a work of art. It was symbolic transposition, not copying of reality [. . .]. In both signs and words, abstraction reflects a gradual adaptation of the motor system of expression to more and more subtly differentiated promptings of the brain. The earliest known paintings do not represent a hunt, a dying animal, or a touching family scene, they are graphic building blocks without any descriptive binder, the support medium of an irretrievably lost oral context.
>
> (p. 190)

Leroi-Gourhan's acute observation – that images are constituted by building blocks that synthesise meaning in various media; be they sound, rock, paper or paint – alludes also to a process of discretisation, inherent in language and technology, that allowed both Jacques Derrida and Bernard Stiegler to question the priority of speech and to address the co-evolution of spoken and written language via the *grammé*, the pro-*gram*, and arche-writing. Discretisation, however, is always already supplemented by synthesis. Metaphoricity attests to this twofoldedness of the construction of meaning that presents itself early on in human evolution. Merlin Donald (1991), for example, argues that gestural communication emerges well before language and believes that 'metaphoric gestures are complex and generative, in that they can be broken down into partial elements and recombined into novel forms' (cited in Modell, 2003, p. 190). In order to understand this better, however, we need to return to the allegorical nature of the human mind.

As we have already seen, Mark Turner (1996) believes that we have certain simple spatial stories at our disposal and that these 'small stories of events in space', themselves 'inventions' of sorts, are projected onto the meanings we imagine and formulate (pp. 13,14). Turner calls these stories 'image schemas' or 'skeletal patterns that recur in our sensory and motor experience', explaining that '[m]otion along a path, bounded interior, balance, and symmetry' are some of these 'image schemas' (p. 16). These simple stories are projected onto the meaning formulated, allowing the emergence of new more complicated stories. However, I strongly believe that these narrative-inventions are also projected to, absorbed and schematised by material technological instances. Could we not, for example, think of the primitive clay pot as instantiating the discrete schema of two hands coming together, forming a cavity, and allowing the holding of liquid? Such a reorientation brings to the fore new ways of understanding technological artefacts; indeed, not simply as extensions of our organs or of our perceptive abilities (see Brey, 2008; Ihde, 1990; McLuhan, 2009; Merleau Ponty, 1964; Verbeek, 2008), but also as material schematisations and metaphorical renderings of discretised schemas of human behaviour. These devices, which I call 'metaphoric machines', allow the emergence of new distinct elements, through the very syntheses that allowed them to emerge as devices, and these are then projected as a whole onto more complicated and synthesised schemas. The term 'metaphoric machines' is quite close to Hayles's notion of 'material metaphor', but my specific formulation suggests that metaphoricity is not a mere transfer of words onto a material object but an ontological process of exteriorisation that first becomes possible via the material embodied and transformative interactions of discrete domains at all levels of life (Kouppanou, 2016).

Through his own methodology, Heidegger clearly showed that he understood the role of schematisation for the syntheses of imagination and time; in fact, by way of celebrating imagination's instantiation in art (*technē*) and in originary/ poetic image, which, in contrast, to technological/representative image, he understood to be functioning as the site of truth. However, we now need to finally assess what aspect of the technological/representative image Heidegger found most restrictive for *alētheia* and for true nearness. Is it materiality, repeatability, reproducibility or even exteriority? As we have already seen, Heidegger appreciated the Van Gogh painting of peasant shoes, namely a material representation of these shoes, believing that such painting can bring us closer to the truth of things; indeed, closer than the actual shoes can. For this reason, we can assert almost with certainty that it is not materiality or repeatability or even reproducibility that Heidegger objects. Of course, it could be argued that art constitutes a unique form of schematisation that is quite distinct from the one manifested with mundane objects; still, according to Heidegger, we can find a site of truth's presencing with the humble jug's outpouring and even with the simple act of passing a bridge. In this respect, we can infer that it is most certainly neither exteriority nor materiality that Heidegger finds threatening in the technological image. What is then his objection, and what is precisely the power he attributes to non originary image?

If we want to begin formulating some kind of answer to this question, we need to turn to what Heidegger understands as the specific characteristics of modern technology – one of them being, technology's tendency to turn things into energy, and the other, technology's propensity to turn nearness into distancelessness. In other words, Heidegger argues that modern media eliminate the possibility of bringing something close, or, to be more specific, of bringing something near without assimilating it into the given, the same, and the predetermined. Furthermore, in *The Age of the World Picture*, Heidegger (1977) addresses modern technology's predisposition to turn everything into a picture. He calls this phenomenon, *getting into the picture*, explaining that through modern technological representations:

> man 'gets into the picture' in precedence over whatever is. But in that man puts himself into the picture in this way, he puts himself into the scene,

i.e., into the open sphere of that which is generally and publicly represented. Therewith man sets himself up as the setting in which whatever is must henceforth set itself forth, must present itself [*sich . . . präsentieren*], i.e, be picture. Man becomes the representative [*der Repräsentant*] of that which is, in the sense of that which has the character of object.

(pp. 131–132)

Modern technology does not simply turn us into stock. It most unmistakably turns us into picture; it schematises us in ways that we ourselves – the very ones making, producing and using pictures – are turned into the made, into the produced and used. This technological effect, which is above all a thinking effect, is the reason behind Heidegger's insistence that technology, similarly to language, is a process of revealing. Technology constitutes the exteriorised metaphorical process that transfers and brings near distinct domains but, according to Heidegger, the modern technological instantiation of this process unfolds by circumventing the human being. Getting into the picture suggests our being embedded in a systematicity that imprisons and assimilates us. But how is this possible? Why do we become objects as a result of perceiving everything as an object? There are two answers to this: First, new media impose their way of perceiving on us. In other words, these technologies are not merely metaphorical renderings of space and time – similarly to the clay pot – but rather constitute metaphorical renderings of perceiving space and time – similarly to the Van Gogh painting. In contrast to the pot, these machines lie closer to the work of art and poetry; they are, above everything, revealing hermeneutic machines allowing us to see things *as* certain things. However, at this point, we need to address a second question, namely: If technology is always already constitutive of our ways of seeing-as, how can modern technologies possibly be any different?

All technologies participate in our hermeneutical processes. This is true. And there is always a technological aspect to *hermeneia*. This is indisputable and supported both by Heidegger and by many contemporary phenomenologists or postphenomelogists like Ihde (1990) and Verbeek (2008). The latter two see technologies in terms of reading. Ihde, for example, explains how the thermometer 'reads' temperature, and how we read the thermometer in order to access the world, while Verbeek explains how the obstetric ultrasound 'embodies a "material interpretation" of reality', offering thus 'a "translation" of what it perceives' (p. 15). These approaches are in accord with Heidegger's take in that they affirm the technological aspect of *hermeneia*. However, as I have argued repeatedly in this book, technologies are at the heart of *hermeneia*, that is, even technologies, which are not supposed to have a hermeneutic aspect. Therefore, I strongly maintain that *hermeneia* is itself a material exterior and embodied metaphorical process unfolding through a twofold process of discretisation and synthesis instantiated through both language and technology. This process allows nearness between gestures, tools, words, and stories and transforms the user. Metaphoricity is imprinted on things and allows things' participation in meaning-making. The human being is but one aspect of this nearing synergy, one-fourth of the fourfold, and therefore it cannot control meaning-making. It does, however, interpret the world limitedly and according to each historical time's possibilities.

Writing is the emblematic technology of this process. It unfolds differently each time, in as much as each system of writing (alphabetic, ideographic, pictographic) and each set of writing implements (pen, typewriter, computer, paper, papyrus, rock, keyboard) allows for a different discretisation and for a distinct figurative spatialisation of thought. Writing becomes the way we come to be as certain beings. The comparison between writing systems, and between their respective possibilities of writing, allows us to see how we are dealing with different ways of seeing the world. With our writing and reading experiences, we often get the chance to experience our perceiving of time and space, while the same holds true for the arts. These technologies do not simply participate, condition, and represent the hermeneutical process, but they rather become metaphoric machines allowing us to see the world as a certain world and also to see ourselves seeing it. Therefore, I believe that we find in Heidegger's disregard of modern technologies the following claim: not only do new media participate in the constitution of *hermeneia* but they also come to dominate it as well. The human being cannot sustain its role as the fourth element of the fourfold; it is rather displaced by automated discretisation, schematisation and interpretation. The reason that this happens is not given by Heidegger, but as my analysis has proposed, this is because the human being is excluded from the process of schematisation while being itself schematised. The human being has at times no understanding of these conditions, rules and processes that construct its own being, as Heidegger (1975) clearly argues in the essay *Building Dwelling Thinking*.

Of course, building, making or producing, were never Heidegger's favourite realms of investigation, since he mostly addressed the tool as finished and used product. With the theory of the fourfold, however, Heidegger, finally, approaches a theorisation of the tool's emergence to being, which is clearly different from the Platonic theorisation of production. With an etymology of the German word for building (*Bauen*) and the verb *buan*, Heidegger (1975) explains in *Building Dwelling Thinking* that to build means to dwell, adding that 'ich bin' means 'I am', and inferring that the way we are as human beings 'is *Buan*, dwelling' (p. 147). Heidegger then distinguishes between two particular connotations of the verb '*bauen*', which, he argues, means not only 'preserving and nurturing', as in agriculture, but also means 'constructing', as in 'ship-building and temple-building' (p. 147). In addition, he retrieves from the 'Old Saxon word wuon' the connotations of staying in place, 'sparing and preserving' that underline an attitude of responsiveness towards things by letting them be instead of submitting them to the human being's will. Heidegger explains then that it is only in this way that we dwell as 'mortals on earth' or as mortals in the fourfold (p. 149). Indeed, existence is linked to building. Moving on to a phenomenological account, Heidegger explains that something like a bridge not only 'gathers' but also 'leads', pointing to the fact that bridges of different historical times construct the experience of being human and of living in time and space differently (p. 147). In this respect, I argue, that bridges, even though not directly linked to *hermeneia* or to reading, constitute metaphoric machines that bring close different domains and aspects of human space and time.

Heidegger's position is quite assertive here in as much as he claims that a bridge is a special type of thing that 'gathers the fourfold in such a way that it allows a site for it' (p. 154). The bridge is a location and only such a thing can 'make space a site' (ibid.). This means that '[t]he location is not already there before the bridge is', but on the contrary location manifests itself 'only by virtue of the bridge' (ibid.). To me, this originary process of spatialisation suggests that technology becomes part of the hermeneutical structure, allowing the very possibility of human space and time through interpretation. In other words, technology and 'building is a distinctive letting-dwell' (p. 159). For this reason, Heidegger says, '[o]nly if we are capable of dwelling, only then can we build' (p. 160). Dwelling is always already a way of being; we dwell as schematised beings, always already transported onto one mode of being or another. Thinking about dwelling is already a thinking originating from this dwelling that corresponds to a certain type of building. This suggests a circularity that is hermeneutic and metaphorical in as much as it reimagines the realities that allow it to exist. The bridge becomes here both literal and metaphorical, and it can be thought as the material metaphor par excellence: the bridge conveys and *transfers* from one place to another, but contrary to what Heidegger initially thought, a bridge does not instantiate a towards-which but a structure that holds together the possibility of here and there, of before and after, and of past and future via the connectedness for which it allows. A bridge is nearness itself, and nearness can be thought through the bridge. In this respect, '[b]uilding and thinking are, each in its own way, inescapable for dwelling' (pp. 160–161). And it is for this reason that mortals 'must ever learn to dwell' (p. 161). The urgency to rethink dwelling comes from technological change. However, is there a possibility that modern technology can make us incapable of imagining new ways of dwelling precisely because it has inserted us into its own structures? Can it be that there are technologies that take away from us the ability to participate in the synthesis and transfiguration of our own time? Can there be metaphoric machines that deprive us from any possible form of participation in the metaphorical process? In order to answer these questions, through the emphasis on the synthesis afforded by originary images, we turn now to a new form of image, namely, digital image.

3. Digital images and metaphors of the digital

Metaphors are quite dominant when digitisation is discussed; from cyberspace to virtual space, from information highway to the World Wide Web, it seems that metaphor has either constructed the experiences we have in the digital sphere or metaphor is our way of understanding these complex experiences and any experiences for that matter. On this point, Werner Kuhn (1996) notes that 'user interfaces use spatial concepts, even if the application domains are not spatial', and that during the process of design, which is called 'spatialization', designers rely on people's lived experiences of space in order to purposefully use spatial metaphors and thus to construe meaningful experiences (p. 1). For Kuhn, this takes place, precisely because '[m]etaphors create ontologies in applications by projecting structure from the source domain. A useful ontology is one that leads to an appropriate work division and effective ways of solving problems' (p. 2).

The technological metaphorisation of nearness is, however, nothing new. Nearness has always been technologically, linguistically, metaphorically, and materially mediated, and, therefore, technology and metaphor cannot be that which defines the specificity of digital tools and digital spaces. Nevertheless, facing Heidegger's objection concerning the impossibility of nearness in a technologically overdetermined era, we need to begin sketching an account of these experiences and especially of the phenomenon he calls *distancelessness*. As with the case of nearness, distancelessness can be understood not as a rendition of a specific preexisting space but as the very constitution of this space and experience, which can be explored with the use of the chain of notions discussed here, namely, imagination, schematisation, image, and metaphor. However, in order to investigate this possibility, let us return to the classical understanding of metaphor, namely, the one resting at the heart of Western metaphysics and the very same that seems to have led us to a certain understanding of virtual spaces in terms of flight of fancy, falsehood and lack of reality. Margaret Wertheim's (2002) account can help us with this discussion. She says:

> In some profound way, cyberspace is *another* place. Unleashed into the Internet, my "location" can no longer be fixed purely in physical space. Just "where" I am when I enter cyberspace is a question yet to be answered, but clearly my position cannot be pinned down to a mathematical location in Euclidian or relativistic space – not with any number of hyperspace extensions!
>
> (p. 300)

What shall we do with these accounts of online experiences? The Heideggerian existential analytic has already established a new way of theorising lived experience via the phenomenon of being-in-the-world and thus we could argue that the experience of space and time – online space and time included – is never merely physical or mathematical. Therefore, if anything at all, the experience of being-online echoes the phenomenological qualities of being-there, being-in and being-near as types of relatedness to the world that by no means require in-one-another-ness but instead denote the technologically mediated constitution of concern. In other words, when space and time are considered as types of relatedness to the world – as they are in Heidegger's analytic – it soon becomes clear that we cannot possibly choose to accept the 'where' of cyberspace or the role of the body or even the lack of materiality as determining factors for the exegesis of being-online. In contrast, we need to think online experiences in reference to being-in-the-world and, therefore, think of them as belonging to the unitary phenomenon of *worldhood* that always already intertwines offline with online experiences and the virtual domains with the supposed 'real' domains. To be more specific, there are at least two ways through which we can proceed to an investigation of digital spaces and digital objects, both of them consistent with Heidegger's work. First, we can follow the example of Heideggerian thinkers, like Michael Eldred (2009), supporting the idea that digital things are mathematically constituted entities and that the 'digital dissolution of beings in progress today' is nothing but 'the consummation of the mathematical casting of being' (p. 12). Eldred explains that this casting of being

> opens up the possibility of calculating with numbers; they are open to λογισμός but at the price (or the advantage) of becoming placeless and positionless. Such a lack of place and position, it seems, characterizes the digital beings which we deal with today. For them, matter in its continuity and fixedness with place becomes indifferent.
>
> (ibid.)

Eldred, of course, is right to identify the abstraction that comes about through mathematical thinking whilst describing distancelessness as an effect of this thinking. This tendency towards abstraction lies, according to Heidegger (1972), at the heart of metaphysics, and it has led to the emergence of the new science of cybernetics – a science that he understood as 'the theory of the steering of the possible planning and arrangement of human labor', able to transform even 'language into an exchange of news' (p. 58). In this respect, cybernetics, or even what we now call digitisation, becomes a structure of total regulation that turns everything into information and every being into a placeless something.

Such a critique, notwithstanding its merits, focuses more on the abstract aspects of digitisation than on our actual dealings with digital objects. Indeed, it could be argued that digital things may be questioned on the ground of their supposed lack of materiality, locality, and substance, but they cannot be questioned on the grounds of their phenomenological presence and meaningfulness. By this position, I do not mean that digital things are necessarily valuable objects; rather, I maintain that they are objects of our intentionality, appearing to us as phenomena synthesised in a meaningful way (Kouppanou & Standish, 2013). Both the 'real' and the 'digital' apple, for example, are objects of our intentionality, and this is even truer for the print and the digital book. The way difference is sustained or eliminated, however, is of critical importance and needs to be investigated phenomenologically.

For this reason, we turn now to a second possible theorisation of digital beings, one pursued by Joohan Kim (2001), who argues that digital beings are either 'informative' or 'executable' (p. 90). The first category consists of things providing 'sensory data', such as texts, images, sounds, videos, etc., and the second category consists of tools, which are ready-to-hand and thus constitute types of 'equipmental contexture' (ibid.). Executable digital beings, such as word processors, Kim adds, are run by computers, providing space for working. These tools 'have no determinable spatiotemporality' because they are 'perfectly duplicable' (pp. 98, 99). However, some digital beings have thing-qualities, such as 'durability', 'degrees of substantiality' – that is, characteristics of colour, size and appearance, 'quasi-bodily presence', the ability to function in terms of reference and also the ability to function in terms of creating sites for interaction (pp. 92–93). Therefore, Kim asserts that digital things, like websites, tend to become inhabitable worlds, commenting that we currently '"dwell" in this "intelligible functionality whole" called the Web' (p. 96).

It thus appears that digital beings have become equipmental structures by means of their propensity to merge with and connect to other tools, things and people and, in fact, despite their presumed lack of materiality. Digital beings are increasingly becoming modes of nearness and of gatheredness. However, with a more careful look, we can see that the nearness they afford is similar to the type of nearness deemed as inauthentic in *Being and Time (BT)*. Digital sites constantly generate the desire for the next projection, the next project, for the next site and the next click, and from this angle they can appear more preoccupied with the towards-which. Despite the fact that I do not agree with Heidegger's distinction between authentic and inauthentic time, we need to wonder, at this point, if digital objects can function like the jug, the painting and the bridge and actually gather the fourfold by allowing the human being to participate in a hermeneutic process that includes both responsiveness and creativity or not.

I have argued here that tools – similarly to words – and technology – similarly to language – can and does function through its own metaphorising powers that produces meaning by gathering domains, projecting stories and transforming the familiar into the unfamiliar. Moreover, these artefactual gatherings, taking the form of languaging or of technological objects, displace meanings: they transform the nearing of domains and the perception of differences. Just like the city bridge gathers the city around it, and the highway bridge expands our reach, technological artefacts have the propensity to gather domains selectively, highlighting and at the same time hiding aspects of experience. However, I do believe that the process of concealing can be translated in terms of automated *hermeneia* in the respect that certain aspects of being need to be hidden, to be negated, or automatically played out, in order for some kind of meaning to be formed. A good way into the investigation of the automated aspects of metaphoricity can be the study of the common clock.

A clock can be thought as a specific metaphoric machine that allows predominantly the thinking of time in terms of space (Kouppanou, 2016). Our experiences of space are, of course, messy and story-like. They include schemas of moving and doing things with our bodies, schemas of things and people coming close to each other, scenarios of walking, resting, passing, grasping, falling, touching, sleeping, waking up and so on. All these schemas, even the simplest one of them, by no means constitutes a simple taking place in pure space. Rather, space and movement in space is always already a mosaic of other complicated spatiotemporal stories involving our acting with things. In other words, the experience of taking place is always messy, involving not only one domain but several ones and, therefore, needs to be filtered and discretised in order to come to us as space and to be projected onto time through spatial exteriorisation onto material surfaces.

For example, we can imagine a time that time is first conceived as the *passage* of the sun. With the clock, time is metaphorised into many different actions and into stories, like the moving of hands and the counting of numbers but, above all, into the *passage* of twelfths and sixtieths. What's more, the traditional round form of the clock allows the projection of the circularity of time and its repeatability. The day following the night and the circular lives of circular celestial bodies are also easily reflected on this structure. All these aspects that the round clock embodies, rest on foundational metaphors that allow the perception of time in terms of space, movement and number. Leroi-Gourhan (1993) comments on the entangled relationship of space-time and technology, asserting that:

> When we say that Moscow is three and a half hours flying time from Paris, we communicate a richer reality that if we alluded to the 2,500 kilometers that separate the two cities: richer because it includes the concept of distance as an experience, just as in the year 1800 it could be said that Lyons was five days away from Paris. By the same token, when we tell the time by looking at a clock we are connecting time to the spatial position of the clock hands.
>
> (p. 315)

With the clock, we are metaphorised beings: transferred and transfigured onto the clock's material nature. Our being in space, and our being-in-the-world, is transcribed as our own being in time that is itself perceived as being at a certain point on a circular and irreversible movement, which nevertheless is guaranteed to present the same time and hour once again. A metaphoric machine, like the clock, enacts metaphoric nearing actions: It gathers our concerns by bringing forth discrete elements of time – like its passing, its circularity and measurement. These discrete elements are resynthesised in new readable unities. This reading, the reading of time, presupposes that the discretisation and the schematisation of time have already taken place and that our own hermeneutic participation is merely a small part of what needs to be done so as a certain understanding of time will come forth. Indeed, '[a] human being needs to read the clock as someone who participates in a metaphorical process, which has already begun, and to experience a type of temporality that opposes existential time' (Kouppanou, 2016, p. 506). As such, automation is engraved in technologies, precisely because these technologies are constituted by smaller, already discretised, and resynthesised schemas.

Technologies of reading and writing seem to be exemplary for instantiating the discretisation and spatialisation of time, based on processes, which are not rediscovered by individual reading but on which the individual relies in order to read. This metaphorical projection of time and space is addressed by Derrida (1997) in terms of spacing, namely, the 'becoming-space of time and the becoming-time of space' (p. 68). Stiegler's (2011a) notion of grammatisation, which is understood as '*the production of tertiary retentions* permitting symbolic fluxes and flows to be discretized and deposited, that is, permitting the spatialization of their temporality, notably in orthothetic forms' similarly alludes to this phenomenon (p. 75).

However, the spatial separation of building blocks, or of meaningful unities, to which we paid heed when describing the figurative behaviour of human beings, is co-constituted through metaphorical projection. Derrida's (1981) discussion in *Positions* opens up the space for such consideration, when he argues that

> [*d*]*ifférance* is the systematic play of differences, of the traces of differences, of the *spacing* by means of which elements are related to each other. This spacing is the simultaneously active and passive (the *a* of *différance* indicates this indecision as concerns activity and passivity, that which cannot be governed by or distributed between the terms of this opposition) production of the intervals without which the 'full' terms would not signify, would not function.
>
> (p. 27)

This type of relatedness and synthesis – as active and passive; indeed, as being the undecidable '*a*' of *différance*, echoes the Kantian schematism to the letter, any letter, not merely '*a*'. In the spacing of *différance*, we can detect some form of synthesis that remains unaccounted for, and, for this reason, Frank Manfred (1989) discusses the notion's incomplete nature, asserting that différance cannot run '*ad infinitum*', infinitely differing and differentiating meaning; 'in the end, it is up to the individual's interpretive and linguistic competence, indeed, even imagination, to decide which term it distinguishes from which other terms in what manner, and with which terms it associates it (metaphorically, metonymically)' (p. 69). In this respect, schematisation, imagination, and synthesis, namely, all the terms with which we began our investigation, return with a vengeance and by means of their binding relationship with discretisation and metaphoricity. Elements need to be temporarily discretised and thus defamiliarised, to be related to each other and resynthesised, to be projected and metaphorically displaced, through the medium of their inscription,[1] and, during a process in which hermeneutics are at times automated. Stiegler argues that the individual user or reader needs to be in position to question 'automatisms that exist in your own relationships to the objects, mathematical objects, physical objects, etc.' (Kouppanou & Stiegler, 2016, 247). However, I believe that a certain degree of automation is unavoidable since the human user cannot learn and cannot know everything at once and every time. Thinking, in its distributed metaphorised form, necessitates automation. The degree to which these automations control the hermeneutical process remains to be deciphered, while the answers to such questions, are ultimately of concern of education. In what comes next, I examine the problem within the frame of specific digital metaphoric machines.

4. The schematisation of nearness: the case of Facebook

A social networking site, like Facebook, appears to be many different things: a tool, an equipmental context, a meeting place, a networking process, a site, and time-synthesis. But do all these definitions constitute an unnecessary pleonasm?

Can something like a tool or a website avoid being a site of connectedness, networking and dwelling? Can a site not be a bridge? This certainly seems to be the case. According to Heidegger (1975), after all: 'a location comes into existence only by virtue of the bridge', which is a 'built thing' (pp. 154,152). In other words, the way something is designed and built is crucial for the kind of experiences it allows. What needs, therefore, to be studied, is the kind of built thing that Facebook is; we need to study its history and architecture. In what follows, I attempt to do precisely that, by turning first to one of Facebook's creator's, namely, Mark Zuckerberg's earlier projects that did not appear to have location- or bridge-like qualities. More specifically, Facemash was a site compiled out of online facebooks (albums) that belonged to Harvard students and that Zuckerberg hacked and used. The site was quite simple, presenting each time two pictures of students and asking users to decide whom they preferred best amongst the two (Hoffman, 2010; Indvik, 2010; Locke, 2007). The purpose, however, of this reference to Facemash is not to turn Zuckerberg himself into a case study but to direct attention towards the great emphasis on likability that this earlier project maintained.

Thefacebook, Facebook's first version, became a sensation among Harvard students, and its popularity increased, as it opened registration to Stanford, Columbia and Yale students. By 1 May 2005, Facebook reached an impressive 800 colleges, and by the 1 September of the same year even high school networks were incorporated. Fasemash's emphasis on locality was very soon replaced by an intense propensity to be connectable to anything else and to enjoy nearness with whatever you preferred (Facebook, 2017a). Right from the start and up until now, Facebook has remained steadfastly faithful to its mission of connecting people, enhancing their bonds and bringing them closer to whatever is relevant to them. As currently expressed in the company's page:

> Facebook's mission is to give people the power to share and make the world more open and connected. People use Facebook to stay connected with friends and family, to discover what's going on in the world, and to share and express what matters to them.
>
> (Facebook, 2017a)

With its February 2017 fourth quarter and full year 2016 results briefs, Facebook reported an impressive count of 1.23 billion daily active users (Facebook, 2017b). Despite this gigantic number of users, and despite the numerous transformations that the site has incorporated, Facebook maintains most of its spatial and nearing characteristics – that is to say, its openness, connectedness and sharing content features – that maintain the site's original emphasis on likability. In order to make this clear, and also in order to show the mechanics of social connectability – as manifested in relation to likability – we now need to take a closer look at this social networking site's (SNS's) specific features.

As already noted, Facebook has transformed dramatically in the last decade but some of its basic features have not. Users get to create their Facebook site called

Page. The notion of 'page' brings to the fore the site's original functionality that allowed each user to log in to their personal site and to navigate manually through other users' websites, as if going through the pages of an album. Currently, owning and using a Facebook account amounts, in summary, to the following actions:

- creating a profile (by posting personal information such as place of residence, work placement, education, date of birth, relationship status, sexual interest, languages spoken, etc.);
- creating updates for your status by answering the question: 'What's on your mind?' Status updates can be in the form of writing or even be multimodal texts consisting of writing, photos, albums, images, links videos and 'live videos'. Updates can also be made by choosing among a variety of activities that describe what you are currently doing or thinking or experiencing and through the doing/feeling feature. In addition, you can also choose to mention a 'life-event', namely, an event that is considered a milestone in your personal and professional life, and lastly, you can add the location of your current activities and the people with whom you are doing them;
- following the News Feed, that is, a summary of all the postings produced by the users of your network, that offers the choice between seeing 'Most Recent Stories' and 'Top Stories';
- friending people, namely, sending and accepting requests, so you can add a new member to your network of friends and be added to different networks;
- commenting on, sharing with, reacting to a post through texts, images and the buttons of 'like', 'dislike', 'love', 'haha', 'wow', 'sad', 'angry', etc.

In what follows, I attempt to look at Facebook's specific organisation as a built-thing; indeed, with an emphasis on the site's own self-descriptive mission and, of course, on metaphoricity.

4.1. Being-there and Facebook

Openness and connectedness can appear at times contradictory, but through the Heideggerian perspective taken in this book, we can infer that openness is in itself a form of inventiveness and a mode of schematisation that allows distinct things to come together by way of new and transformative connections. The possibility of such synthesis is, after all, what Heidegger understands as nearness and originary/poetic image, whilst he perceives its impossibility – or what he calls *distancelessness* – as modern technology's most threatening characteristic. We are then about to consider Facebook as a digital image and investigate the kind of nearness and synthesis it affords. Bearing in mind that all these distance-related notions refer to the experience of time, and to the way that time is formulated through metaphoric schematisation, I will now attempt to understand Facebook as mediator, site of nearness and metaphoric machine that projects spatiotemporal scenarios of being onto its own discrete schemas.

Metaphor, as discussed here, connotes the projection of an overt story onto a covert one, the selective correspondences of messy domains, and the nearing that allows the emergence of new discrete elements from domains previously thought as transparent and as known totalities. As I have already proposed, there cannot be absolutely known domains, and similarly there is no pure non-mediated and non-spatial temporality. Indeed, even Turner's simple spatial stories are never merely spatial; any form of movement in space is always already a spatiotemporal scenario, a dynamic interactivity and a synthetic nearing. In this respect, we cannot have recourse to any purely temporal or purely spatial stories, which are then supposedly projected onto Facebook's own domains. In contrast, following the line of thinking opened up by Heideggerian phenomenology, we need to sketch the source of our 'overt' story in terms of *Dasein, being-in-the-world* and the complicated and multifaceted story that these structures presuppose. Without a doubt, if there is going to be any kind of transfer online, it is going to be a projection of *being-there* onto an online configuration that is itself understood always already as being-there.

Being-there is predominantly temporal in Heidegger, being itself a selective projection of a complex schema of lived experience onto Heidegger's phenomenological phrasing, but I have already argued here that being-there cannot and, according to later Heidegger, is not merely temporal. It is rather spatiotemporal, embodied and material. Since this multiplicity and dispersion is our reality, we can then infer that every time an online *being-there* instantiates *being-there*, it brings to the fore a selected number of aspects of being-there and leaves others in the background. With a first look, then, it seems that Facebook attracts mostly temporally oriented aspects of being, which are then projected onto the site's spatial features. But, which temporal aspects these are, and how they are retemporalised by users, awaits for clarification.

Being-there is, as we have seen, an essentially complicated manifold, linked to being-in-the-world, which is itself constituted by the worldhood of the world and by the relation of being-in that is not experienced in terms of in-one-anotherness but in terms of inhabitation, dwelling, thrownness, and nearness. *Dasein*, Heidegger argues, is characterised by its propensity to bring things near, that is, things defined by their own propensity to be serviceable, useable, available, and near. A digital thing like Facebook has, as Kim's analysis (2001) has already shown, almost all of these characteristics. It can even said to be ready-to-hand, indeed, ready-in-the-physical-hand, through the constitutive and initiating schematisation of the body that allows connection to any-*thing*. Facebook moreover appears to have an enhanced propensity to refer to other things and thus to establish networks. If we want, however, to know which discrete building blocks Facebook industrialises the most, we need to look at the existential makeup of any *being-there*.

According to Heidegger (2008), *Dasein* is always already found in some mood or another. Moods are states of being: in other words, they constitute ways through which the world comes to us and they make up the phenomenal content upon which we have hard time to reflect. Moods like fear, boredom, anxiety, and joy constitute primordial ways of knowing, not accessible by cognition: we are just the way that we are, already fallen in one state or another, always already attuned to something. As Heidegger says, '[*e*]*xistentially, a state-of-mind implies a disclosive submission to the world, out of which we can encounter something that matters to us*' (p. 177/138). As a matter of fact, one of Heidegger's aims, as stated in *BT*, is to elevate this indecipherable content to the level of phenomenological visibility and to thus open it up to hermeneutic investigation. As he says:

> Phenomenological Interpretation must make it possible for Dasein itself to disclose things primordially; it must, as it were, let Dasein interpret itself. Such Interpretation takes part in this disclosure only in order to raise to a conceptual level the phenomenal content of what has been disclosed, and to do so existentially.
>
> (p. 179/140)

This primordial content is not easily raised to the level of phenomenological visibility, and it is not easily put into words. If we think about it, however, this is precisely what Facebook demands from us when it asks us to update our status. Indeed, during the first years of its founding, and more specifically on the 1 September 2004, Facebook launched its Wall feature (Facebook, 2017a), which constituted a surface inviting users to post an update concerning their lives by responding to the question, 'What's on your mind?', and offering the choice to write or upload a text, a photo, a link, a video or a question. In this case, the projected notion of *writing on the Wall* suggested that online *being-there* was to be thought and experienced more in terms of wandering around a modern city and of gazing at its readable walls and less in terms of reading a book.

As of 22 September 2011, the Wall was replaced by Timeline (Facebook, 2017a). This notion brings forth different connotations, since 'timeline' hints at the graphic representation of historic time, chronology, narrative order, and the selective projection of being-there as temporal flow. Facebook's Timeline, however, does not represent a time that has already taken place, but rather constitutes a process that allows time to take place through this specific exteriorisation. Timeline is an object that, similarly to the clock, spatialises time and allows the user to retemporalise it. Still, contrary to the clock, Facebook's emphasis is not on the passage and on the passing of time, but rather on the synthesis of the present. Timeline, after all, appears to be a stream of now-moments presenting themselves to individual users. These now-moments are isolated memories awaiting for their schematisation, and this schematisation is able to synthesise a greater present. Indeed, Facebook constitutes a pool of memories, ours and others': I can draw some of these memories and I can constitute my own present by responding to them. The notion of otherness is quite important here, and it comes more into play with the Facebook feature, News Feed, namely, the representation of a summary of postings made by the users of my network and presented to me as that which is most current, relevant and closer to my concerns through the options of 'recent' or of 'most popular' stories.

Drawing from the Husserlian and Kantian background, which allowed the constitution of that which Heidegger called imagination, and from Derrida's and Stiegler's respective readings, we need to investigate now the degrees of passivity and of spontaneity for which some Facebook features potentially allow. Anticipation becomes, as we have seen, projection onto the future by drawing memories from the past, allowing thus for the synthesis of the now. In other words, being-there has to draw from past memories in order to draw some kind of now. Memories are kept in me and are also kept in tools and devices. Through the use of devices, I let certain memories to participate in my synthesis of time. A tool is therefore 'pros-thesis'; it is always already set in front of me, not only spatially but also temporally: it guides my future by supplementing myself and by constructing possibilities of my becoming (see Stiegler, 1998). In our time and age, however, Stiegler (2011a, 2011b) argues, that it is the programming and culture industries that form our time, imposing their selection criteria on our attention and thus constituting not only the milieu in which we find ourselves but also our being-there. In order to understand this process, we need to turn now to another instance of online mediated nearness and then return to the time-synthesis offered by Facebook's Timeline and by News Feed. Search engines are a good entry point into this investigation.

4.2. Search engines and nearness

Search engines work on the premise that a disparate and abundant domain of online information needs to be filtered in order for every individual user to receive information that is important for and relevant specifically to them. Such access, however, needs to be determined based on specific processes and criteria. Google's search engine, for example, works through the following three steps: First, its automated programs, called Googlebots or spiders, access websites either by crawling from link to link or via URL, and then they download these pages in Google's browsers. After that, Google's indexer 'sorts every word on every page and stores the resulting index of words in a huge database' so that the query processor 'compares your search query to the index and recommends the documents that it considers most relevant' (Blachman & Peek, 2007). Relevance is determined by different variables, one of them being PageRank, an agglomeration of more than a hundred selection criteria, 'including the popularity of the page, the position and size of the search terms within the page, and the proximity of the search terms to one another on the page' (Blachman & Peek, 2007). As a patented algorithm, Google's PageRank cannot be revealed, but we know that it 'gives more priority to pages that have search terms near each other and in the same order as the query' (Blachman & Peek, 2007). In this way, PageRank emphasises the popularity of pages, the proximity, the measured presence, the quality of content and the quality of links connected to the page, being, however, only one among Google's 200 variables responsible for judging content (Google, n.d.).

It thus appear that Google's aspiration to find important pages for specific users is very much defined by set standards that maintain and also construct a specific image of what type of knowledge is out there and ultimately of what is the *out-there*. What is noteworthy is that in this case, nearness is understood more as a photographic spatial depiction of terms and of words and less as an innovative negotiation of differences, and as synthesis. Nearness is counted more as a measurable presence and less as an implied proximity, which can possibly disturb the familiar and allow for the alien. This is, however, not the only way users are affected, since the set criteriology that constitutes the engine's indexing and judging of relevancy, is also affecting their experience of online space and time. For this reason, Kitchin (2017) states that '[m]any of the most important algorithms that people encounter on a regular basis and which (re)shape how they perform tasks or the services they receive are created in environments that are not open to scrutiny and their source code is hidden inside impenetrable executable files' (p. 20). Similarly, Lucas D. Introna and Helen Nissenbaum (2000) argue that 'not only are most users unaware of these particular biases [of algorithms], they seem also to be unaware that they are unaware' (p. 176). Otherwise put, the representation of information by search engines is a schematisation that hides the fact that it is a form of synthesis by excluding and selecting information.

Online retentions are produced by each and every search and by each and every user. Search engines rely on these retentions in order to customise and to personalise the results of queries or even the queries themselves. This type of personalisation finds its most common instantiation in Google's 'autocomplete' function that generates queries by completing search words. In this way, the retentions that I leave online become the criteria of my search: they synthesise my protentions and construct my anticipation.[2] Despite possible time gains, this process that the 'autocomplete' simply represents, can lock the user into a temporality in which the has-been does not simply contribute to the formation of anticipation but fully dominates it: it suspends possibilities of difference, deferral and deviation, leaving the searcher doomed to return to the same, to an identity routinely reinforced by recurrent retentions, and to a time formed by an external selection of memories. The traces users leave behind and the filters that decide as to that which is to be retained shape the nature of our world, affecting therefore the *polis* and indeed politics in the grander sense, and raising ethical and political questions. 'Memory is,' as Stiegler (2009) puts it, 'always the object of a politics, of a criteriology by which it selects the events to be retained' (p. 9). Now, if we also consider that memory is the basis of imagination and of time, we can begin to think search engines in terms of their being our exterior organs of synthesis, knowing and being, positioning us in pictures we have not ourselves sketched.

When we enter the picture, we become part of this world, and we are subjected to its limits. This effect intensifies in the digital sphere, forcing Robert Luke (2003) to invoke the *panoptikon*, arguing that: 'while we watch stock quotes filter through our portals, we are tracked in this watching in a panoptic, pan *info* con becomes the archive even as we create it, add to it, participate within it' (p. 335). In the same vein, Stiegler (2009) comments that:

> Internet users are invited to produce tags, keywords, indexations and annotations of all kinds for this 'new screen,' which becomes a collaborative effort,

what one calls Web 2.0 and which constitutes the participative architecture of an infrastructure itself based on cloud computing. This has led to an age of the bottom-up production of metadata, which in turn constitutes a radical novelty in the history of humanity.

(p. 52)

The traces users leave behind and the filters that determine that which is to be retained create the nature and the limits of our world. When we enter the picture, we schematise and are schematised by that which comes near. We ultimately become that which is near.

Let us, however, return now to Facebook's Timeline that seems to instantiate the opposite of this process, focusing, in contrast, on each individual user's posts and memories. Indeed, one could even argue that Facebook's Timeline could potentially become our own stream of consciousness, our evolving historicity, our time in the world, our story and archive of memories. But then again, we need to pay attention once again to what it is that Facebook is asking from us when it necessitates the update of our status.

4.3. Facebook's metaphoric power

What is on our mind, towards what is our consciousness directed, and with what we are ourselves concerned, relates to phenomenology's most important discovery: that we are always directed towards something. This directedness or concern is precisely phenomenology's problem. This concern is, as we saw, formed by that which has passed. However, that which has passed, our memories, are not always and simply ours. As inheritors of our communities and part of their *there*, we share memories that we have not lived but are at our disposal. In this respect, search engines become the mediators of our access to these memories through the very criteria they impose on us. In opposition to this search engine propensity, it could be argued that Facebook is focused on the user, allowing one's own choosing and creative synthesis of time. With choices between 'feeling, celebrating, watching, reading, getting, thinking about, listening to, meeting, looking for, travelling to, playing, drinking, remembering, and, of course, with the open invitation to write whatever you like and to choose among emotions, actions, books, food, and so on, Facebook seems more concerned with each individual user's specific concerns than with a given algorithm. By inviting us to say what is on our minds, namely, what constitutes our moods and states of being, Facebook asks us to consciously contribute memories that can potentially act as the selection criteria of what is to come near specifically to us. Writing appears to be such a process, a play of presences and absences, attracting not simply a response from the reader but also allowing the constitution of the writer. The exteriorisation afforded by writing suggests a series of absences: First, we have the absence of an interior self always already in the process of making something and of being made into something exterior. Then we have the second absence of the self, who as the writer or the sender of the written text, is absent from their writings, whilst they 'continue to produce effects beyond his presence and beyond the present actuality of his meaning, that is, beyond his life itself' and finally, we have the 'absence of the addressee', having always already participated in the formation of meaning of the written text (Derrida, 1982, p. 313).

Writing on Facebook plays with all these absences, while a certain negation – that is, a negation of absence, which according to Derrida 'belongs to the structure of all writing' comes to the fore (ibid.). Someone writing on Facebook and, in fact, writing about that which is on their mind seems to inhabit a middle space between self-expression that parallels private journaling, and self-exteriorisation that parallels epistolary communication. In both cases, some kind of addressee is construed through the writer's/sender's performing of the writing self. This writing, and this inscription of memory, becomes our own selection criterion for our further postings and for others' postings as well. Facebook offers a constant supply of memories, made out of postings and of responses to these postings. However, in contrast to a journal entry, which might never get a response, or to letter writing, which instantiates a certain temporal distance between sending and receiving, a Facebook posting necessitates the immediate and self-evident response of others in order for this posting to claim some kind of presence and degree of reality. Indeed, postings that do not receive immediate responses do not appear in other users' News Feeds and thus never gain presence or the potentiality to become a selection criterion for their responses and for further actions.

News Feed was a feature introduced on the 5 September 2006, constituting a compilation of new information, appearing on a user's network (Facebook, 2017a). Since the selection of what is to come close, through the News Feed, is determined by the rhythm of response, writing on Facebook appeared to promote a temporality of immediacy and presence. As of 9 February 2009, this emphasis was firmly solidified through the Like button that encouraged users to push it whenever they enjoyed something they saw (Facebook, 2017a). In this way, the amount of Likes a posting received became the measurable evidence of some thing's worth, that is, worth to be seen and to gain presence on other users' New Feeds. The further emphasis on likability was, however, not the only result of this change, since liking or not liking something became a way of orienting users' writing towards the increase of likability, and also a way of codifying the users' responses. Despite, these apparent limitations, it could be argued, that this mechanism encouraged users' own choosing of responses instead of the machine's choosing: users choose what they consider important and allow it to come close. Users of Facebook had and have until now one more choice: they get to decide if the site shows them the most recent or the most popular stories of their network, and, in this light, Facebook appears to be less determined by the company's agenda and more by individual users. These are not, however, the sole criteria determining Facebook's nearness, since, similarly, to search engines, Facebook employs its own algorithm that decides on the presence of the elements appearing on each user's News Feed. Because of the vastness of this information and the finitude of human perception, new content needs to be summarised, and the possibility of such summary came to be through the EdgeRank, Facebook's algorithm, used up until 2013.

According to Taina Bucher (2012), there were three main factors according to which relevancy was calculated by EdgeRank. The first axis was *affinity*, determined in accord with the degree and kind of interaction between two users. The second was the *weight* that each action (called Edge) was given by Facebook. A comment, for example, was given more weight than a 'Like', and, of course, chatting with someone was considered as even more important. The last axis was the *time decay*. The weight of each Edge was counted according to presence and to quantity and not according to actual content. A hate comment, for example, had equal weight with a comment of encouragement as long as it had equal duration. Finally, the weight depended 'on the internal incentives' of Facebook (p. 1169). This is still quite true, since, at any moment, Facebook is attempting to promote some new feature or other, be it 'Moments', 'Videos' or some new modality, and therefore gives these features more visibility.

Since 2013, this algorithm has, however, changed, and Facebook is now using a more complicated ranking system 'based on machine learning' that takes into account more than '100,000 individual weights' (McGee, 2013). Moreover, in 2015, a decisive new filter was added to Facebook's ranking system, having to do with the amount of 'Likes' a story received and also with the time spent reading it, whilst in 2017 a similar filter was added, concerning the viewing time and the actual completion of a video (Bapna & Park, 2017). These changes seem to indicate Facebook's aspiration to establish more valid assumptions concerning people's interests, namely, ones that focus more on the individual user instead of the quantity of interactions between people, but again these mechanisms appear to be processes of discretisation of nearness, through the pushing of buttons and the calculation of engagement, that attempt to assess and replicate the users' already schematised – largely by the SNS's own mechanisms – nearness.

In the light of the previous discussion, it appears that the question, 'What is on your mind?', which is, in fact, a question concerning our state of being and about our definite thrownness, is answered by us, but the responses to this answer, and to our own online performances, are greatly defined by Facebook that becomes our own organ of nearness. This does not mean that Facebook extends one of our bodily organs, but that it descritises and resynthesises an already complicated existential condition. Indeed, as Bucher (2017) argues, 'Facebook's algorithms become part of 'force-relations' and are generative of different experiences, moods and sensations' (p. 39). It thus becomes clear that in exteriorising ourselves as partial elements of Facebook's architectonics, we sometimes gain visibility and we sometimes do not. However, by not being-present on other users' News Feeds, we become invisible, almost less real, or even inexistent. That is why, Bucher (2012) asserts that contrary to the panopticon that enforces a rule of total visibility, Facebook's mechanics 'install visibility in a much more unstable fashion: one is never totally seen or particularly deprived of a seeing capacity' (p. 1171).

4.4. The prevalence of nowness in Facebook

Our investigation has shown up to this point that a digital thing like Facebook can be understood through the mathematical casting of being that transforms everything, every human action and meaningful hermeneutic exchange, into a discrete and measurable element, used as a selection criterion for our ongoing actions of nearness. This process, however, is also present in all metaphorical and hermeneutic processes, and, for this reason, I have claimed that the calculability of digital things and of digital actions is always already supplemented by meaningful practices that turn even trivial choices like pushing the 'Like' button into constitutive acts of our presence. Facebook's mechanisms do not affect our time solely by choosing memories on which we rely in order to synthesise time, but the site's specific features appear to dominate the synthesising process. What's more, Facebook's emphasis on newness appears to formulate time as a linear passing focused on what is peaking or trending and is thus presented as an autonomous now. With each new post, users are thrown into new circles of nowness, lasting from minutes to days, but the aspiration – even when a post is successful – is that another new and equally successful post will be written once again. In other words, if something is not trending now, its user ceases to be relevant and visible to collective consciousness. The configuration of this nowness, we could even argue, is nothing different from an algorithmic illusion appearing as a certain something; it is the manifestation of time *as* time that either includes you or it does not, and this is precisely the reason that users experience this time so intensely and personally.

The emphasis on nowness is currently supplemented by other Facebook features that appear to contribute to different types of time-synthesis. The 'On This Day' feature, for example, brings to the fore an event that you lived in the past, the 'Year in Review' feature constitutes a selection of the user's past year memories synthesised in the form of a video-narrative and so is the 'Celebrate Your Friendship' feature that celebrates the anniversary of two users' online connection by reporting the history of this friendship and statistical elements of the users' interactions such as amounts of 'Likes' they exchange etc. The user can either choose to post this remembering and, therefore, activate it by making it available to other people or not. Temporal linearity is also challenged by Facebook's most spatial features that either allow you to mention your location at the time of your posting, through the 'Check in' feature, or mention the people you are with, via the 'Tag' feature. What's more, with the 2014 'Nearby Friends' and the 'Travelling' features you can see which one of your friends is close and the destinations of their journeys (Facebook, 2017a). Still, as already noted, even these actions of returning to the past, and of spatial situatedness, quickly dissolve into sharply delineated now-moments constituted by the degree of immediate response and of likability they draw from other users.

5. Conclusion

In attempting to bring this discussion to a close, we can argue that Facebook's openness is schematised through its means of connectability that do not merely bring people and things closer to each other but also synthesise their inscribed memories. One user's past memories are the other's mediated anticipation of the future. Every post appears to be a memory that the user's consciousness can employ to synthesise time, and these memories often taking the form of short stories or snapshots – videos, photos, texts, etc. – instantiate a particular life-moment, which is then projected onto a particular middle space resting uncomfortably between the private and the public sphere, enacting the user's temporality.

These stories are essentially smaller schemas that reveal some aspects of the user's life and at the same time conceal others. They are also posed as questions that demand some kind of response. At a personal level, this anticipation can be elating or soul-wrenching, depending on the amount of 'Likes' one's content receives and on the durability that their online presence gains, but on an existential level, this suggests a type of human temporality conditioned by the constant need of newness, which can be fed by an autonomous system. At the level of communal space, however, this emphasis on what is peaking and trending can potentially allow transformative reorganisations of space and of time that may lead to differentiations in political performances. The emphasis on the users' nowness can, indeed, contribute to the maintenance of a trend that would normally be excluded by traditional media as was supposedly the case with the Arab Spring (Khondker, 2011).

Facebook's metaphorising power is, however, nothing fixed; it is an evolving process, and it will continue to locate discrete elements of human behaviour and to synthesise them in ways it sees fit. These metaphoric processes should not be taken lightly, since Facebook has become a vast pool of the users' exteriorisations and performances that creates what is now called big data. Big data is information gathered online, and it can be analysed in order to reveal patterns of behaviour of either individual users or communities of users and thus can offer accurate depictions of these users' profiles. Indeed, according to Kosinski, Wang, Lakkaraju and Leskovec (2016), liking a particular singer can tell something about our musical taste, but correlations between seemingly random 'Likes' can allow the detection of patterns and offer accurate information about the 'psychodemographic user profiles' (p. 494). This possibility underlines the connections between the social media's metaphoric devices and the political sphere, since profiles can be targeted in such a way as to potentially affect the social and the political behaviour of users by offering to them partial individualised information concerning a product or even a political issue. In this respect, users do not get the picture but merely get to become part of the picture. This, in turn, permits the assimilation of users into a specific picture that becomes their own version of the world.

As users we constantly become part of the very images we are using; we enter the picture by allowing our complicated being-in-the-world to be discretised, mostly without our knowledge or consent and to be resynthesised and reoriented towards that which is deemed important to us by others. In this way, digitisation seems to be a form of nearness that we do not choose but do perform. These technologies decide as to the nearness and farness of things and as to the way our spatiotemporalities will be formulated (*Bilden*). They thus become dwelling sites. However, the kind of dwelling they provide is not that of the house but that of the workshop; they are gatherings in some way, but they are constantly projecting us onto the next project and the next 'Like'.

As my analysis has shown, we cannot simply say that digital tools turn everything into mathematical abstract entities or that they do not offer meaningful hermeneutic experiences. Indeed, their way of interchanging between these modes of existence might as well be the reason that they become all the more powerful while entertaining the possibility that their metaphoric power will become so automated that it ends being perceptible by the user. What this means for education, however, namely, the very process that attempts to select and to project memories that students need to resynthesise, is discussed in the next chapter.

Notes

1 Thinkers like Roman Jakobson have underscored the fact that metaphoricity exists in language in terms of sign choice and in terms of sign association, taking place *in absentia* at the level of paradigmatic relations, regulated by metaphorical substitution, and *in presentia* at the level of syntagmatic relations, regulated by the metonymical combination of signs (see Jakobson & Halle, 1956).
2 Accordingly, Google (2017) informs us that Autocomplete takes into account '[t]he terms you're typing. Relevant searches you've done in the past (if you're signed in to your Google Account and have Web & App Activity turned on). What other people are searching for, including Trending stories'.

References

Bapna, A., & Park, S. (2017). News Feed FYI: Updating How We Account for Video Completion Rates. *Facebook Newsroom*. Retrieved March 10, 2017, from http://newsroom.fb.com/news/2017/01/news-feed-fyi-updating-how-we-account-for-video-completion-rates/

Blachman, N., & Peek, J. (2007). How Google Works. *Google Guide*. Retrieved January 22, 2013, from www.googleguide.com/google_works.html

Black, M. (1962). *Models and Metaphors: Studies in Language and Philosophy*. Ithaca: Cornell University Press.

Brey, P. (2008). The Computer as Cognitive Artifact and Simulator of Worlds. In A. Briggle, P. Brey & K. Waelbers (Eds.), *Current Issues in Computing and Philosophy* (91–102). Amsterdam: IOS Press.

Bucher, T. (2012). Want to Be on the Top? Algorithmic Power and the Threat of Invisibility on Facebook. *New Media & Society*, *14*(7), 1164–1180.

Bucher, T. (2017). The Algorithmic Imaginary: Exploring the Ordinary Affects of Facebook Algorithms. *Information, Communication & Society*, *20*(1), 30–44.

Derrida, J. (1981). *Positions* (A. Bass, Trans.). Chicago, IL: University of Chicago Press.

Derrida, J. (1982). *Margins of Philosophy* (A. Bass, Trans.). Chicago, IL: University Of Chicago Press.

Derrida, J. (1997). *Of Grammatology* (G. C. Spivak, Trans.). Baltimore and London: The John Hopkins University Press.

Facebook. (2017a). Company Info. *Facebook Newsroom*. Retrieved Janurary 10, 2017, from http://newsroom.fb.com/company-info/

Facebook. (2017b). Facebook Reports Fourth Quarter and Full Year 2016 Results. *Facebook Investor Relations*. Retrieved March 10, 2017, from https://investor.

fb.com/investor-news/press-release-details/2017/Facebook-Reports-Fourth-Quarter-and-Full-Year-2016-Results/default.aspx

Fauconnier, G. (2001). Conceptual Blending. In N. J. Smelser & P. B. Baltes (Eds.), *International Encyclopedia of the Social & Behavioral Sciences* (pp. 2495–2498). Oxford: Elsevier Science Ltd.

Fauconnier, G., & Turner, M. (2002). *The Way We Think: Conceptual Blending and the Mind's Hidden Complexities.* New York, NY: Basic Books.

Frank, M. (1989). *What Is Neostructuralism?* (S. Wilke & R. Gray, Trans.). Minneapolis, MN: University of Minnesota Press.

Google. (2017). Search Help: Search Using Autocomplete. *Google Help.* Retrieved March 10, 2017, from https://support.google.com/websearch/answer/106230?hl=en

Google. (n.d.). How Seach Works: Algorithms. *Google Inside Search.* Retrieved March 1, 2017, from www.google.com/insidesearch/howsearchworks/algorithms.html

Hayles, K. (2002). *Writing Machines.* Cambridge and London: MIT Press.

Heidegger, M. (1972). *On Time and Being* (J. Stambaugh, Trans.). New York and London: Harper & Row.

Heidegger, M. (1975). Building Dwelling Thinking. In A. Hofstadter (Ed.), *Poetry, Language, Thought* (pp. 141–160). New York: Harper and Row.

Heidegger, M. (2008). *Being and Time* (J. Macquarrie & E. Robinson, Trans.). Oxford: Blackwell.

Hoffman, C. (2010). The Battle for Facebook. *Rolling Stone.* Retrieved January 10, 2017, from www.rollingstone.com/culture/news/the-battle-for-facebook-20100915

Ihde, D. (1990). *Technology and the Lifeworld: From Garden to Earth.* Bloomington: Indiana University Press.

Indvik, L. (2010). Mark Zuckerberg's Infamous FaceMash.com for Sale. *CNN.* Retrieved January 10, 2017, from http://edition.cnn.com/2010/TECH/social.media/10/06/facemash.mashable/

Introna, L. D., & Nissenbaum, H. (2000). Shaping the Web: Why the Politics of Search Engines Matter. *Information Society, 16*(3), 169–186.

Jakobson, R., & Halle, M. (1956). *Fundamentals of Language.* The Hague: Mouton & Co 's-Gravenhage.

Khondker, H. H. (2011). Role of the New Media in the Arab Spring. *Globalizations, 8*(5), 675–679.

Kim, J. (2001). Phenomenology of Digital-Being. *Human Studies, 24,* 87–111.

Kitchin, R. (2017). Thinking Critically about and Researching Algorithms. *Information, Communication & Society, 20*(1), 14–29.

Kosinski, M., Wang, Y., Lakkaraju, H., & Leskovec, J. (2016). Mining Big Data to Extract Patterns and Predict Real-Life Outcomes. *Psychological Methods, 21*(4), 493–506.

Kouppanou, A. (2016). Texts as Metaphoric Machines and the Challenge of the Digital. *Educational Theory and Philosophy, 66*(4), 499–518.

Kouppanou, A., & Standish, P. (2013). Ethics, Phenomenology, and Ontology. In S. Price, C. Jewitt & B. Brown (Eds.), *Sage Handbook of Digital Technology Research* (pp. 102–116). London: Sage.

Kouppanou, A., & Stiegler, B. (2016). ". . . Einstein's Most Rational Dimension of Noetic Life and the Teddy Bear . . ." An Interview with Bernard Stiegler on Childhood, Education and the Digital. *Studies in Philosophy and Education, 35*(3), 241–249.

Kuhn, W. (1996). *Handling Data Spatially: Spatializing User Interfaces.* Paper presented at the 7th International Symposium on Spatial Data Handling, SDH'96, Advances in GIS Research II, Delft, The Netherlands.

Lakoff, G., & Johnson, M. (1980). *Metaphors We Live By.* Chicago and London: University of Chicago Press.

Leroi-Gourhan, A. (1993). *Gesture and Speech* (A. B. Berger, Trans.). Cambridge, MA, London, England: The MIT Press.

Locke, L. (2007). The Future of Facebook. *TIME.* Retrieved January 10, 2017, from http://content.time.com/time/business/article/0,8599,1644040,00.html

McGee, M. (2013). Edge Rank Is Dead: Facebook's News Feed Algorithm Now Has Close to 100K Weight Factors. *Marketing Land.* Retrieved March 10, 2017, from http://marketingland.com/edgerank-is-dead-facebooks-news-feed-algorithm-now-has-close-to-100k-weight-factors-55908

McLuhan, M. (2009). *Understanding Media: The Extensions of Man.* London and New York: Routledge.

Merleau Ponty, M. (1964). *The Primacy of Perception: And Other Essays on Phenomenological Psychology.* Evanston, IL: Northwestern University Press.

Michael, E. (2009). *The Digital Cast of Being: Metaphysics, Mathematics, Cartesianism, Cybernetics, Capitalism, Communication.* Germany: Ontos Verlag.

Modell, A. H. (2003). *Imagination and the Meaningful Brain.* Cambridge, MA: MIT Press.

Ricœur, P. (2004). *The Rule of Metaphor: The Creation of Meaning in Language* (R. Czerny, K. McLauughlin & J. Costello, Trans.). London and New York: Roudtledge Classics.

Robert, L. (2003). Signal Event Context: Trace Technologies of the habit@online. *Educational Philosophy and Theory, 35*(3), 333–348.

Stiegler, B. (2009). *Technics and Time 2: Disorientation* (S. Barker, Trans.). Stanford, CA: Stanford University Press.

Stiegler, B. (2011a). *Decadence of Industrial Democracies* (D. Ross & S. Arnold, Trans.). Cambridge: Polity Press.

Stiegler, B. (2011b). *Technics and Time, 3: Cinematic Time and the Question of Malaise* (S. Barker, Trans.). Stanford, CA: Stanford University Press.

Turner, M. (1996). *The Literary Mind.* Oxford: Oxford University Press.

Verbeek, P.-P. (2008). Obstetric Ultrasound and the Technological Mediation of Morality: A Postphenomenological Analysis. *Human Studies, 31*(1), 11–26.

Wertheim, M. (2002). The Pearly Gates of Cyberspace: A History of Space from Dante to the Internet. In N. Spiller (Ed.), *Cyber Reader: Critical Writings for the Digital Era* (298–303). London: Phaidon.

8 Education as *Bildung* and as formation according to image

1. Introduction

At the beginning of this book, I presented the educational rhetoric emanating from Heidegger's discourse of Enframing as, on the one hand, a philosophy of education articulating the most appropriate concerns about learning and teaching – that is, concerns about technology's effects on thinking, and, on the other hand, I criticised this same theory for being an inadequate philosophy of technology, neither able to explain the interconnectedness of language and technology nor able to pay heed to new technological instances like digital tools and to the particular ways that these tools affect being and thinking. Despite these shortcomings, the Enframing discourse brings to the fore the intricate relationship that binds together technology and education, and, in fact, in ways in which other educational theorisations of technology do not. These latter perspectives are mostly drawn from instrumentalism, namely, the theory of technology that perceives tools as neutral instruments that cannot influence the actions for which they are used (Feenberg, 2006). These perspectives are roughly the following:

a   The *common-sense* perspective maintaining the neutrality of technological means. David Lewin (2013) explains that '[t]he idea that technological devices are neutral certainly appeals to common sense. It would seem bizarre to imagine that devices could have subjectivity, desires, or could determine their own ends. Rather we assume that devices are subject to *our* will – to human purposes' (p. 2). Such a perspective assigns education the role of the doer. Education chooses certain instruments in order to serve the aims and purposes it aspires to achieve, while these purposes are considered to be in no way affected by the tools used. Software and material devices and also courses and teaching approaches like Information and Communication Technology, Media Education and Digital Literacy, are thus often treated as the means that will eventually provide the necessary skills for survival in the real world. The fact, however, that these technologies incorporate specific ideologies and are always already synthesising-metaphoric machines that impose their modalities of synthesis on the user is usually not explicitly addressed. An aspect of this phenomenon, which contributes to the uncritical application even of the most critical approaches, is highlighted by David Buckingham (2003) who comments that '[c]omputers are aggressively marketed to parents and teachers as an educational medium – indeed, as *the* indispensable educational tool for the modern world' (p. 174). However, Buckingham adds that:

> computers are largely seen here as delivery mechanisms – as neutral means of accessing 'information' that will somehow automatically bring about learning. 'Wiring up' schools is often seen to produce immediate benefits, irrespective of how these technologies are actually used.
>
> (ibid.)

Buckingham's discussion points to a rather confusing position: On the one hand, technologies are understood as neutral means lacking intrinsic modes and values and are therefore uncritically embedded in schools, and on the other hand, technologies, which have never been designed with any specific learning purpose in mind, are perceived as something that essentially procures learning and is therefore intrinsically knowing-effecting.

b   The *enhancement* perspective resonates with the common sense perspective, in the respect that it takes technology to be a 'tool' that improves human activities and especially production. It is subsequently inferred that if technologies used 'out there' increase productivity, then these technologies are bound to have the same effect 'in here', namely, in education. Technology enhances learning. Enhancement, however, is usually limitedly perceived in terms of efficiency, and therefore this belief increases the demand for demonstrable results proving that learning has taken place. This approach, therefore, usually fails to consider the complicated coupling of technologies and learners, even though numerous mismatches have been noted. Andrew Dillon and Ralph Gabbard (1998), for example, argue that 'the evidence does not support the use of most hypermedia applications where the goal is to increase learner comprehension', even though this is what practitioners fervently attempt to do (p. 334). In addition, theorists like Martin Oliver (2012) comment that, in fields like educational technology, '[t]echnology itself is seldom considered, being treated instead as "natural" or given. [. . .] The consequence of this is that technology is treated as if it will cause learning – and when it does not, there is no clear explanation of why' (p. 31). Finally, Neil Selwyn (2011) asserts that:

> Educational technology is an essentially 'positive project.' Most people working in this area are driven by an underlying belief that digital technologies are – in some way – capable of improving education. This mindset is evident, for example, in the recent tendency to refer to 'technology enhanced learning' or before this to 'computer-assisted learning' – descriptions that both leave little doubt over the inherent connection between technology and the improvement of learning and teaching. As such, the *de facto* role of the educational technologist is understood to be one of finding ways to make these technology-based

improvements happen and – to coin a phrase often used in the field – to 'harness the power of technology'.

(p. 713)

These are not, however, the only assumptions drawn from the instrumental perspective, since apart from the neutral and the confusingly learning-effecting role that they ascribe to technology, they also assert that the world and education are two isolated domains. In this way, the effects that technologies already have on practitioners, beliefs and students are never set under examination.

c   The *constructivist* perspective, inspired by thinkers like John Dewey, Jean Piaget and Lev Vygotsky, is quite distinct from the aforementioned positions in the respect that it supports the idea that learning constitutes a relation between the learners and their environments (see for example Dewey, 1997; Piaget, 1999; Vygotsky, 1978). As such, learning should not be about following instructions but about the experiential ways through which the learner constructs knowledge and things. Since the learner is the one constructing the knowledge, however, technology is merely understood as a neutral tool. Therefore, this perspective, and contrary to the other positions, acknowledges the importance of technology for learning – in fact, technology becomes indispensable for learning, but it also conceptualises the student 'as the active subject that makes sense of a passive learning environment' (Ream & Ream, 2005, p. 583).[1] In this respect, certain forms of constructivism reinforce the subject-object, and, the world-education, dualisms, and in consequence the belief in the instrumental nature of technology.

d   Lastly, an approach that admits the necessity to know about the specificity of technology comes from the traditional take of *cognitive science* that perceives human learning to be some form of emulation of computer functioning. This belief is evident in remarks made by theorists like Robert B. Kozma (1987) who argues that '[t]o be effective, a tool for learning must closely parallel the learning process; and the computer, as an information processor, could hardly be better suited for this' (cited in Friesen & Feenberg, 2007). This approach, however, makes a huge inference: it takes a tool that is itself a selective instantiation of certain thinking functions and elevates it to the level of an unchanging reality – indeed, turns it into a theory that supposedly represents human thinking (Kouppanou, 2011). The educational translation of this is the following: if a certain tool simulates human learning; learning processes should essentially incorporate this tool. Its incorporation will produce learning in any case. This phenomenon is explained by the 'tools to theory heuristic', inspired by Gerd Gigerenzer (2002) – supporting that technologies, and their wide use, condition the construction of theories that describe how the human mind works (cited in Friesen, 2010). Indeed, Friesen (2010) explains that

> in applied disciplines like educational technology and human-computer interaction, technology plays two important but conflicting roles. It first

operates heuristically to explain complex mental phenomena; it is then designed and developed explicitly as a tool for facilitating and developing these same complex mental processes.

(p. 83)

Learning about learning, then, is an arduous task – especially because the tool is always already involved in the possibility of performing and understanding thinking, and it is always already conditioning the human mind (Kouppanou, 2011). Where does this leave us?

The first three conceptualisations of technology show that understanding technology as a neutral means constitutes a weak perspective that leads education to rather awkward efforts and practices. Similarly, the last approach unveils technology's conditioning effects on thinking either through educational practices or through the theorisations of thinking that limitedly perceive thinking as the very tool they use. This circularity reveals a deeper need to understand what it is that technologies are doing and what it is that we are doing when we use them. Indeed, even instrumental theories of technologies appear to assume that technology affects thinking in some way, but what we need to ascertain is the precise way; any effect of technology on thinking should not be understood as a desired learning outcome, and any desired learning outcome should not be thought independently from the tool used. Heidegger's discourse of Enframing, which is a kind of an essentialist approach, agrees in part with the last perspective presented here, but the equation of technology with one tendency will prevent us from detecting the specificity of new technologies. There is thus the need for a perspective of philosophy of technology in education that 'promises the possibility of an understanding of technology that may be important not only to public policy but also in helping to conceptualise intellectual approaches to the study of technology and, indeed, to shaping new fields of knowledge and research' (Peters, 2006, p. 96). In what comes next, I summarise the perspective discussed and applied in this book.

2. Technology and *Bildung*: the perspective of metaphoricity

The perspective taken in this book could serve the role of a new philosophy of technology in education, as it attempts to both unveil important facets of human situatedness, such as nearness, imagination and metaphoricity, and correspondingly investigate these existential types of relatedness in connection to specific technologies and domains that involve them. The simultaneous study of human experiences that include technologies and of technologies that include human beings cannot offer clear-cut domains, but rather domains that always already include and interact with each other. This methodology is necessitated through the very structure of metaphoricity that is incidentally able to explain the reasons behind the presence of so many technological metaphors in theories that conceptualise the mind.

Indeed, Norman Friesen (2010) argues that we have at times come to understand the mind through the metaphor of the clock, the camera and the computer. However, I believe that this is so because machines selectively incorporate and metaphorise certain schemas of human experience. In consequence, these metaphoric machines bring to the fore an understanding of human experience that is then projected onto theoretical structures of understanding thinking – like educational discourses. Taking this possibility into account, we can assume that the potential contribution of philosophy of technology to education is not to be limitedly confined to the ways we use tools. To the contrary, such contribution would be capable of proposing a clarification of technology's nature in connection to an illumination of the human being and of thinking. It could also show how individuals, communities and institutions are conditioned by technology, and how education could potentially condition the human being according to its own particular aims.

A part of this discussion has already begun unfolding. Katherine Hayles (2007), for example, has explained how modern technologies promote the user's constant shift of attention and have thus produced a cognitive mode that she calls 'hyper attention'. This mode of attention entails the engagement with multiple tasks, the need for more stimulation and the intolerance to boredom (p. 187). On the opposite side, lies 'deep attention', which we usually associate with humanities, and 'is characterized by concentrating on a single object for long periods' (ibid.). Hayles sees ADHD (Attention Deficit Hyperactivity Disorder) as the extreme, but now common, manifestation of hyper attention. Relying on research from 'brain imaging studies', she infers that the use of different media causes synaptogenesis to proceed in different ways. To put this in other terms, different technologies – the book, the video game, the web – wire our brains differently in as much as 'the brain's synaptic connections are coevolving with an environment in which media consumption is a dominant factor' (p. 192).

In this respect, it seems that technology is literally formation and literally *Bildung*: it is the formation of our minds; indeed, as Nietzsche asserts: 'Our writing tools are working on our thoughts' (Kittler, 1999, p. 200). Tools have different affordances that extract different types of embodied reactions, rhythms and modes of perception from us, and these define the limits of that which can be anticipated and imagined. In this context, a brain wired according to the conditioning that a video game environment inflicts will not be able to engage with cognitive tasks that demand concentration over long periods of time such as literary reading. This realisation, however, points to another fact; indeed, that what counts when it comes to digital technologies is not simply the content or the memories constituting our selection criteria, but also the type of connectedness that these memories afford, and that they in turn metaphorise our modes of attention, thinking and imagination. In fact, we can understand thinking itself as the response to the tools we use and at the same time as that which constructs our site of being, our individual and collective *there*: tools synthesise our receptivity, constitute the origin that allows the transfer of meaning, and schematise metaphorically what comes near.

For this reason, Stiegler (2010) calls these technologies 'psychotechnics' (p. 93). The term can be a little misleading, bringing to the fore binaries of body and mind or body and soul, questioned by both Heidegger and Stiegler. However, Stiegler's (2010) own argument in *Taking Care of Youth and the Generations* builds upon Hayles's analysis and describes how psychotechnics ultimately form the brain, pointing in this way to the unbreakable unity of the psychosomatic. As we saw, data from neuroscience research supports these views, offering images that give a clear picture of what it is to get into the picture, that is, to interiorise the functionality of tools at the level of the synapses. The tools we use ultimately form a circuit, binding the individual with the environment and thus creating a kind of hermeneutic circle that determines what is to come near. So the situation, at least for Stiegler and for Hayles, appears like this: the brain because of its plasticity is without form, and thus it is always already undergoing a process of formation. This realisation, however, creates a responsibility for increased awareness about that which causes the formation.

'Human formation' usually constitutes the rendition of *Bildung* in English. Indeed, education has been traditionally thought to be endowed with this responsibility: to form the human being. However, if technology conditions the environment in which we are embedded, and if technology wires our *there* in the ways that we have been describing, can education still claim that it is a process of formation? *Bildung* means molding oneself according to an image, a picture or vision – a *Bild* in German. However, Sven Erik Nordenbo (2002) points out that within the educational context, we can say that

> a person has acquired *Bildung* only if he or she has assisted actively in its formation or development. In other words, in the educational context, the concept of *Bildung* contains a reference to an active core in the person who is *gebildet* [educated, cultured].
>
> (p. 314)

In this respect, technology can be the cause of formation. Technology itself is a type of *Bildung*, but if the individual has not in some way contributed to this process of formation, then this formation is not really *Bildung*. The individual needs to be allowed to individuate themselves, and this means that the individual needs to be aware of that which is causing the forming. In our current context, this translates as a need for literacies – media literacy, digital literacy, and multiliteracies – that investigate critically and go far beyond the superficial explorations of media representations. This also demands a type of knowing that allows us not only to understand the technological *Bild* (image) that causes the formation and probably obstructs educational *Bildung*, but also that allows us to come to know how we come to know in general.

With this book, I attempted to participate in this discussion by discussing first Plato's *Allegory of the Cave*. In this text the images (shadows) produced by the fire are contrasted to the real images (ideas) of objects found outside of the cave and illuminated by the sun. In his own reading, Heidegger maintains the distinction between artificiality and nature, even though he presents *Bildung* in a quite similar way to technical production, that is, to the very model of thinking that he considers responsible for the construction of Western metaphysics (Kouppanou, 2014). Heidegger's reading of the Platonic text does not allow him to pay enough attention to the role of the fire, that is, the device that produces artificial images. The liberated prisoner attempting to leave the cave faces for the first time the source of their world; one could even say the source of their reality – the fire. The realisation that there is another reality, namely, the one of the physical realm, the sunlight and ultimately the sun, does not come separately from the awareness of the fire's presence and role. The sun, in fact, can only be understood and related to the prisoner in terms of metaphor and thus in terms of analogy to the fire. The prisoner's newly acquired knowledge allows them to proceed to the other reality. However, the shadows continue to be part of our real world, that is, part of that which allows the liberated human being to understand both fire and sun. For Plato, the sun is the ultimate reality. For Heidegger, there is clearly no such ideal origin, but clearly there exists the possibility of an *alētheic* nearness to the world, challenged by certain images. This nearness is interconnected to the nearness of the fire. The fire needs to be studied not as a derivative way of knowing but as a constitutive factor for the state of being human, and this responsibility befalls education as it is the formative and the transformational force that metaphorises the human being's individuation.

3. Philosophy of education and philosophy of technology coming together

Contributing to the much needed dialogue between philosophy of technology and philosophy of education, I addressed here the possibility that both technology and education are processes of human formation. Through a reconceptualisation of metaphoricity as an always already material and embodied movement allowing the discretisation of domains, their proximity, the projection of memories and the figuration of stories, I discussed the possibility for the user to be transfigured and indeed be figured-out in line with figures of speech and figures of technics encountered in the world. Technology is a profound way through which human beings are formed. Each historical context, relying on its own technological means, produces different individuals. In this respect, technology contributes to the constitution of the *already-there* in which any individual finds themselves always already thrown and always already formed. Education's nature, however, as a process of formation (*Bildung*), which precisely aspires to be a process of *paideia* and indeed *paideia* through certain beliefs about that which is true and worthwhile, needs to be a different type of formation, relying on different types of images.

For Heidegger, technology – even, when it turns our existence into distancelessness – is always already a technology of situatedness: Facebook, for example, orients and mediates us – as every tool does – towards a specific concern, no matter how short-lived this concern may be. Education, conversely, necessitates the '*atopon eikona*', the image of no place and the place of nowhere. This is, nonetheless, a different nowhere from the nowhere-everywhere of digital spaces; it is the nowhere of becoming and emerging; namely, the nowhere of an unpredictable not-determined and even risky synthesis that has always already begun. It is the nowhere of creative absence and of responsive imagination, and finally it is the nowhere of nearness. Such type of nearness necessitates decision, postponement and critical self-examination, whilst technologies like Facebook demand imme-diacy, predetermination and newness, and, for this reason, René Arcilla (2002) wonders: 'How seriously and critically can one examine oneself if one is pre-vented, by the nature of the online experience, from examining what supports that self-examination, namely, its medium?' (p. 463).

Technology precedes, proposes and forms. Education follows while being endowed with the burden of making decisions concerning the one being formed and indeed by attempting to make it possible for the one being formed to make their own decisions concerning their individual forming. However, how, if at all, is this process of decision-making possible? Let us attempt to think this through. Technology is a form of *mathésis*, of learning or of what Stiegler calls epiphylo-genesis, namely, a recapitulation of past knowledge, acquired through its respec-tive exteriorisation on and through us. Think about the technology of reading, for example, and the fact that reading becomes possible only after a certain rewir-ing of the brain has already taken place. As Maryanne Wolf (2008) states:

> Unlike [. . .] vision and speech, which *are* genetically organised, reading has no direct genetic programme passing it on to future generations. [. . .] This is part of what makes reading – and any cultural invention – different from other processes, and why it does not come as naturally to our children as vision or spoken language, which are preprogrammed.
>
> (p. 11)

Technologies, and most certainly technologies of reading and writing, form us, and form especially those who are most formable, namely, children. Technolo-gies found in a child's environment – and by technologies I refer to anything, from the pacifier to the tablet – constitute modes of synthesis through which a child is formed as a child of a specific technologically co-constituted milieu. As such, childhood is perhaps the most important stage of human development: it is a stage during which the absence of a human origin is most intensely felt, since learning becomes the child's mode of connectedness, whereas the alreadyness of the world is first constituted as alreadyness (Kouppanou & Stiegler, 2016). Heidegger was never explicitly concerned with childhood or with learning or even with the constitution of the ready-to-hand. When these matters are, how-ever, finally addressed in his work, nearness emerges as a process and a matter of learning. The human being learns what is near, often by becoming what is near through the differential possibilities offered by languaging and technologi-cal tools. A child alternates between possibilities of relatedness to technology and possibilities of relatedness to speech, which themselves participate in the emergence of language and technology. Both materiality and embodiment have important roles to play for this process by allowing the exteriorisation of the self through and onto the world, tools and images.

Discussions concerning the derivative nature of images, either understood as imitations that lack materiality or copies that re-*present* an already formed presence, are not at all relevant here. Images are always already exteriorised schemas and syntheses that invite the user to use them by subsequently allowing their selves to be schematised and synthesised accordingly. Currently children's worlds are dominated by technologically produced images, which might not appear as images at all, and these images condition the way children think and experience their time and space. This kind of conditioning becomes the child's process of individuation even before the child enters any stage of formal schooling, and it is, for this reason, that we need to be aware of each technology's effects on thinking and on imagination. We also need to be aware of the metaphors we use to describe each technology, since these metaphors become themselves explanatory devices not only of specific technologies but also of the way we think or we perceive the world in general.

In this respect, education takes place after a certain process of technological formation has already taken place and indeed simultaneously to technological formation's multifaceted unfolding. Education's role, however, is, for this reason, all the more important. Education conditions modes of knowing and aspires to allow the scaffolding of knowledge through technological means. During this process, some processes of thinking need to be or to become automated so that more spacetime is given to more complicated processes or to the ones that the student needs to learn. In this respect, education should seek to use technologies while identifying the ways technologies use and metaphorise learners. Even, however, in cases, that technologies present themselves to us as black boxes, we should remember that the human being is perhaps the black box par excellence and that our theoretical involvement with technologies should proceed through a twofold concentration on both the human and the technical components of that which constitutes our situatedness.

Technologies constitute processes of synthesis and metaphoricity transferred onto the brain. This transfer suggests a re-organisation of interiority that is able to function through and as new technological modalities. Such a transformation involves a degree of passiveness in the respect that learners do not know the kind of formation that technologies inflict on their own organisation, and they likewise lack knowledge concerning the exact ways that these technologies were designed and produced. In other words, learning by means of the technological supplement takes place interchangeably between stages of active participation, during which we learn to do something with technologies, and passive ones, during which these technologies are somehow imprinted in our hands, bodies, brains, time and space. In both cases, however, technology is μόρφωσης[2] – it is education and formation. What's more, technologies necessitate automation in their own specific unfolding, and through their combined networked unfolding with other technologies, precisely because of human finitude. The human being cannot do and cannot learn everything, and it cannot learn it at once. There needs to be a distribution of tasks between the human being and its environment. Concerns, skills and even desires need to be disseminated and performed by machines, that is, performed in non-originary ways so that the human being itself will be able to experience the originary and the spontaneous nearness of things. Automated metaphoric processes become the substratum of open metaphoric and truly creative ones, just like multiplication becomes an automated mathematical process that is potentially combined with innovative and speculative mathematical thinking. Automated processes are needed for originary and creative transformations, and open metaphoric processes are themselves regulated to a certain degree. Think, for example, of a process like writing a poem and consider how grammatical and syntactic rules, or even what linguists call paradigmatic and syntagmatic relations, are expected to take place in order for new schematisations to emerge.

Heidegger did not see the interaction between originary and non-originary image and certainly did not explicitly address their differences, despite the fact, that such distinction was clearly very important for his thought. In contrast, he underlined that modern technology's tendency to eliminate hermeneutic openness and imaginative connectedness is able to creep up in every human activity, to take over and to make the human interpretive participation redundant for the hermeneutic task. Returning to the example of poetry writing, we can imagine, a situation during which the configuring of a poem is transformed from a spontaneous responsiveness to an automated process of combining discrete linguistic, phonetic and thematic patterns, a process that might or might not involve the use of software. A similar scenario can also be discussed when considering a student's reflection concerning philosophical thinking through the internet. Think, for example, of a young person writing a philosophical paper on the possible link between critical thinking and education, taking the following steps: First, they type critical thinking in Google's search bar. Then, they choose the first website offered to them, probably Wikipedia's page on critical thinking. Third, they navigate Wikipedia's page, moving from link to link and paying the expected amount of attention to etymology, definition and critical thinking's procedural stages. Fourth, they read Wikipedia's section on the Deweyan connection between critical thinking and education. Finally, they argue that critical thinking is such and such and reach the conclusion that it is certainly good to have critical thinking in education; indeed, without ever questioning the notion, its constitution, its connectedness to other realms, and ultimately without being themselves critically concerned or even engaged with critical thinking. Now, if we think that search engines, like Google, can potentially take away from us the very process of questioning, which Heidegger considered as the most defining human characteristic, then we can begin detecting processes that eliminate hermeneutical freedom.

Similarly to Heidegger, Bernard Stiegler is particularly concerned with the possibility of the elimination of human mediation and detects in the rhythms of the televisual system, and especially in its capacity to simultaneously register and broadcast information, a specific 'technical calculation of time that is so rapid it evades the scope of phenomenological consciousness' (Hansen, 2004). In the same vein, Max Horkheimer and Theodor W. Adorno (2006) have described the industrialisation of imagination in the following terms:

The active contribution which Kantian schematism still expected of subjects –
that they should, from the first, relate sensuous multiplicity to fundamental
concepts – is denied to the subject by industry. It purveys schematism as
its first service to the customer. According to Kantian schematism, a secret
mechanism within the psyche preformed immediate data to fit them into the
system of pure reason. That secret has now been unravelled. [. . .] For the
consumer there is nothing left to classify, since the classification has already
been preempted by the schematism of production.

(p. 44)

Heidegger, Stiegler, Horkheimer and Adorno address the problem of *alētheia*,
différance and imagination respectively, and all of them see the specific technological instances of their own time to be impregnated with the possibility of
these notions' impossibility. This leads us to the realisation that the specificity
of the technologies we use and possibly their tendency to eliminate difference
is precisely the question of our time. In *Echographies of Television* (Derrida &
Stiegler, 2002), Stiegler is pressing Derrida on this matter – the former affirm-
ing that technology being prior to différance allows for the possibility of différance's undoing – and the latter arguing that différance is itself the condition
of the possibility of time, indeed any time, mediated by any technology. As Mark
B. N. Hansen (2004) puts it, '[b]ecause he thinks that *différance* simply is more
originary than technics, Derrida can take for granted the possibility for a critical
relationship to teletechnologies' whereas for Stiegler this is not the case, since
the 'possibility for transcendence is itself transductively correlated with technics'.

Bernard Stiegler's and Martin Heidegger's critiques of technology converge in
this respect: They both detect the danger of technology's outside being eclipsed.
For Stiegler, this translates not simply into an automation of *hermeneia* but also
into the complete lack of knowledge or of access to that which conditions *hermeneia*. Time is always already synthesised into unities – the discrete constitutive schemas of which remain hidden. Similarly, for Heidegger, distancelessness
becomes not simply the everywhere and nowhere of experience but also the too
near that negates the possibility of a type of nearness that is essential for the consideration of that which is coming near. In this light, Stiegler's deconstructive
reading of both Heidegger and Derrida reaches a possible impasse, becoming
itself another type of theoretical totality that similarly to Enframing constitutes a
critical observation about technological tendencies and at the same time becomes
an impediment to the critical phenomenological investigation of technological
specificity. This realisation, however, brings a whole new problematic to the
fore; in fact, one regarding the ways with which these philosophers understand
the co-constitutive relationship of the human and the technical, the ways they
understand these realms separately, and the ways they perceive technology to
be taking over this relationship. Hansen (2004), for example, does not believe
that technical time conflates with human time and argues that other aspects of
human synthesis should be taken into consideration, when the eclipse of human
synthesis is discussed. I could not agree more with this position; in point of

fact, when the emphasis is shifted on metaphoricity, other elements of synthesis come to the fore, having to do with the co-constitutive relationship of media, the human body, the materiality of technological artefacts and the languaging nature of things. All these aspects function not merely as forces of differentiation of memories but also as modes of synthesis and imagination. This opening towards synthesis is perhaps the reason that allows Stiegler (2002) to focus on the digital image or on what he calls '*l' image de synthése*', as the realm in which subjective intervention, choice, and production can still take place (p. 148). Such focus on synthesis can go both ways, illuminating not merely the openness of digital image but also the openness of human synthesis, which is, as I have argued, richer than what Stiegler proposes.

As we saw, digital objects can be thought as existing interchangeably through the processes of mathematical reasoning and hermeneutic connectedness. Indeed, both mathematical reasoning and creative metaphoricity participate in the design of these objects, whilst their use can be closed and hermeneutical. However, the foundational Heideggerian insight, namely, that tools are always already found in networks, suggests that even if a certain tool limits hermeneutical openness, it can still lead to other tools and networks that do allow for it. This possibility of connectedness is always already part of an open structure. Returning to the example of the poem, we can think of an author writing poetry after having in any case automatised the skills of alphabetic writing, longhand writing, keyboarding and, indeed, the skill of finding and reading poetry in the print and digital realms. In this way, we can easily see that a lot of syntheses need to be predetermined and automated in order for a truly open metaphoric process to take place and that metaphoricity as originary image and true nearness relies on this series of automatic processes. Another example comes from Facebook itself and the way its automated technics of nearness, that is, its reminders of birthday celebrations and of events taking place near a certain user, can lead to open and unexpected proximities online or offline. Finally, in the same vein, we can think of the way representational space and automated nearness, offered by GPS devices, are able to afford experiences of nearness to specific locations and places. We can then infer that there are various stages of the hermeneutic-nearing process, which are either spontaneous or automated, and of which we have various degrees of knowledge and access. Even in the case of alphabetic writing, we have to accept, according to Derrida, that we live in a state of 'relative illeteracy' (p. 59). This, however, opens up further questions for education and its responsibility – ones with strong political and economic connotations: If, in fact, we will always be in a state of relative illeteracy, we need to be at least concerned with the ways that new technological tools relate to and form our milieu and to not negotiate our rights to have access to these black boxes either ourselves or through trusting intermidiaries. In this light, education should be concerned both with knowledge and with the political claim of the access to this knowledge.

With my analysis so far, I have shown that I can understand the prevailing tendencies that both Heidegger and Stiegler detect in modern technologies, but I cannot easily accept the possibility of the elimination of human participation. Indeed, even if it was possible for teletechnologies to completely enforce their own selection criteria on our imaginations, these technologies' inherent connectedness to other media – older and new, makes them amenable to change and to singular appropriation. What's more, this connectability refers not simply to forms of arranging discrete elements next to each other but also to choices and syntheses allowed by the technical realm and performed by the human, and these syntheses are complicated, embodied, dispersed and multifaceted. We thus need to turn our look towards both aspects of synthesis, namely, the human and the technical, and approach creatively both the technological and the human black boxes. Accordingly, education needs to be always trying to achieve the unachievable, dealing not only with a formation that has already begun but also with a kind of interpretation that can at any time become more critical, creative and imaginative.

4. Conclusion

The twofold methodology undertaken here is a good first step towards a new way of thinking about technology and about education. We need, however, to continue investigating the nature of human thinking through phenomenological and through empirical investigations of new technologies that offer access to the way we allow the world to be inscribed in us from the most abstract level to the most specific one of brain imaging. Second, we need to attempt to think schooling in new ways: instead of thinking what needs to be taught in terms of selecting memories, and thus, in terms of facts, information and subjects, we need to begin thinking about teaching in terms of metaphoricity, modes of synthesis and metaphoric schemas. Metaphoricity as a basic human propensity can be imagined and cultivated through its different instantiations in our mathematical, scientific, poetic, literary, conceptual and emotional lives. In a time that more and more of human skills are becoming automated, we need to begin imagining what the new human being's needs and desires would or should be. Going into the third point, education needs to participate in the dialogue concerning the post-human, not in terms of abandonment of the human, but in terms of forsaking a dream of an origin that never was. Fourth, education needs to be involved in the political discussions concerning technology, society and the economy. Stiegler, contrary to Heidegger, and precisely because he accepts that humanity is always engaged and in fact produced by a process of technologisation, pays attention to the way economic and market interests affect this process, and we can assert that even if not all users are capable of understanding technological automations, they need to be capable of understanding how economic interests that employ these automations work. In contrast to Stiegler, Heidegger sought to sketch a separation between humanity and technology that partly led to his tragic political mistakes. This is because the understanding of technology in isolation from the political and the economic questions that concern it can further lead to the obfuscation of technosociopolitical matters.

Technology re-enters and reshapes the complicated nexus of connections that constitute society, indeed, in such a way that it is difficult to deduce the nature of its influence. Technology co-constructs the relational sub-stratum, which is our home and familiarity. However, technology is always already a networked, multi-faceted and dispersed reality responding to the human networked, multifaceted and dispersed existence. Concerns as humanly and as technologically mediated spatiotemporalities are instantiated in multiple ways, and, for this reason, I think that it is wrong to focus on the tendencies of modern technology – like Enfram-ing, immediacy, connectivity and digitisation – and consider them in isolation from previous tendencies and structures of technology and also in isolation from other forms of human dispersion like embodiment, language and materiality that continue to interact with what we consider to be hegemonic. It is, for this reason, that I believe that education's task is difficult but not impossible: the co-constitutive metaphoric process in which technology participates will not be so easily dominated and controlled by a single tendency, whereas technology's interaction with other realms will continue creating opportunities for the consti-tution of new metaphoric and critical schemas.

Notes

1 This is not to say that these thinkers and especially Dewey's line of thought cannot initiate rich discussions about technology. However, the instrument is not often directly addressed in the educational appropriation of these theories.
2 Μόρφωσις is another Greek word for education that connotes formation: it is derived from *morphe* (form) and connotes the taking of a shape through education.

References

Arcilla, R. V. (2002). Modernising Media or Modernist Medium? The Struggle for Liberal Learning in Our Information Age. *Journal of Philosophy of Education*, *36*(3), 457–465.

Buckingham, D. (2003). *Media Education: Literacy, Learning and Contemporary Culture*. Cambridge: Polity Press.

Derrida, J., & Stiegler, B. (2002). *Echographies of Television* (J. Bajorek, Trans.). Cambridge: Polity.

Dewey, J. (1997). *Democracy and Education: An Introduction to the Philosophy of Education*. New York, NY: Free Press.

Dillon, A., & Gabbard, R. (1998). Hypermedia as an Educational Technology: A Review of the Quantitative Research Literature on Learner Comprehension, Con-trol, and Style. *Review of Educational Research*, *68*(3), 322–349.

Feenberg, A. (2006). *Questioning Technology*. London: Routledge.

Friesen, N. (2010). Mind and Machine: Ethical and Epistemological Implications for Research. *AI & Society*, *25*(1), 83–92.

Friesen, N., & Feenberg, A. (2007). 'Ed Tech in Reverse': Information Technologies and the Cognitive Revolution. *Educational Philosophy and Theory*, *39*(7), 720–736.

Hansen, M. (2004). 'Realtime Synthesis' and the Différance of the Body: Techno-cultural Studies in the Wake of Deconstruction. *Culture Machine*, *6*. Retrieved March 10, 2017, from https://www.culturemachine.net/index.php/cm/article/view/9/8

Hayles, K. (2007). Hyper and Deep Attention: The Generational Divide in Cognitive Modes. *Profession*, *1*, 187–199.

Horkheimer, M., & Adorno, T. W. (2006). The Culture Industry: Enlightenment as Mass Deception. In M. G. Durham & D. M. Kellner (Eds.), *Media and Cultural Studies: Keyworks* (pp. 41–72). Malden, MA: Blackwell Publishing.

Kittler, F. A. (1999). *Gramophone, Film, Typewriter* (G. Winthrop-Young & M. Wutz, Trans.). Stanford, CA: Stanford University Press.

Kouppanou, A. (2011). Learning about Learning through Technology. *Bajo Balabra: Revista de Filosofía, 6*, 117–126.

Kouppanou, A. (2014). Imagining Imagination and Bildung in the Age of the Digitized World Picture. In M. Papastephanou, T. Strand & A. Pirrie (Eds.), *Philosophy as a Lived Experience: Navigating through Dichotomies of Thought and Action* (pp. 77–89). Berlin: LIT verlag.

Kouppanou, A., & Stiegler, B. (2016). ". . . Einstein's Most Rational Dimension of Noetic Life and the Teddy Bear . . ." An Interview with Bernard Stiegler on Childhood, Education and the Digital. *Studies in Philosophy and Education, 35*(3), 241–249.

Lewin, D. (2013, 22–24 March). *No Place for Wisdom: Technological Thinking and the Erosion of Phronesis.* Paper presented at the Philosophy of Education Society of Great Britain, Annual Conference, New College, Oxford.

Nordenbo, S. E. (2002). Bildung and the Thinking of Bildung. *Journal of Philosophy of Education, 36*(3), 341–352.

Oliver, M. (2012). Learning Technology: Theorising the Tools We Study. *British Journal of Educational Technology, 44*(1), 31–43.

Peters, A. M. (2006). Towards Philosophy of Technology in Education: Mapping the Field. In J. Weiss, J. Nolan, J. Hunsinger & P. Trifonas (Eds.), *The International Handbook of Virtual Learning Environments* (pp. 95–116). Netherlands: Springer.

Piaget, J. (1999). *The Construction of Reality in the Child.* Oxford: Routledge.

Ream, T. C., & Ream, T. W. (2005). From Low-Lying Roofs to Towering Spires: Toward a Heideggerian Understanding of Learning Environments. *Educational Philosophy and Theory, 37*(4), 585–597.

Selwyn, N. (2011). Editorial: In Praise of Pessimism-the Need for Negativity in Educational Technology. *British Journal of Educational Technology, 42*(5), 713–718.

Stiegler, B. (2002). The Discrete Image (J. Bajorek, Trans.). In J. Derrida & B. Stiegler (Eds.), *Echographies of Television* (145–163). Cambridge: Polity Press.

Stiegler, B. (2010). *Taking Care of Youth and the Generations.* Stanford, CA: Stanford University Press.

Vygotsky, L. S. (1978). *Mind in society: The development of higher psychological processes.* Cambridge, MA: Harvard University Press.

Wolf, M. (2008). *Proust and the Squid: The Story and Science of the Reading Brain.* New York, NY: HarperCollins Publishers.

Conclusion

Throughout the history of philosophy, there has always been a quest for the missing link, the third thing, the mediator that brings the human being near to the world. This thing has been called by many names – idea, understanding, analogy, being, imagination, representation, différance – and each of these theoretical formulations has accordingly assigned different roles to the human being, the world, thinking, image, word, sign, tool, and the medium. Heidegger's reconceptualisation of the third thing is conducted through his investigation of time and imagination and iterated as poetic image and nearness.

Heidegger's conception of time gravitates towards schematisation and synthesis and can be best thought in terms of a conversation with Kant and with Aristotle, focusing not so much on the nature of representation, as Derrida would have it, but mostly on time's constitution as imaginative nearing. The reinscription of imagination as time is presented in many different ways in Heidegger, but the poetic/originary image and nearness are perhaps the most pertinent ones. This originary image is not originary in terms of pre-existing or of being some kind of referent/signified to a supposed derivative technological representation/image, but in terms of giving time *as* certain time, that is, by maintaining both the elements of passivity and of acceptance, by allowing things to be *as* they are and to emerge in transformative crossings and as new hermeneutic unities. Indeed, this is how the present comes to be, namely, through the nearing of the past to a possible future. The present emerges out of the selection and of the projection of past memories onto expectation, whilst transforming the future by bringing it close. Time comes to *be* by means of this nearing.

Nearness develops into Heidegger's third thing: It is a process that brings things near in a circular motion; it brings the known closer to the unknown, but it juxtaposes these domains in such ways as to shed light on the alien aspects of the familiar and on the familiar aspects of the uncanny. Therefore, it can be reconceptualised as a movement of metaphoricity. Heidegger, of course, resists such possibility by irreversibly considering metaphor as constitutive of metaphysics and as the essence of technology. This possibility, however, namely, the possibility of metaphor being the essence of metaphysics and by extension of technology, which is according to Heidegger a revealing mode as well, is quite indicative of the possibility to reconceptualise metaphoricity in terms of emergence but not essentially in terms of the metaphysical or the technological. Heidegger has described many processes under the guise of this metaphoricity – *aesthesis, hermeneia*, meditative thinking and fourfold being perhaps the most enticing of these – and has articulated a philosophy enfolded in an irrevocably metaphorical phrasing, which he considers to be not metaphorical at all. All these loose ends offer the ground to think metaphoricity, in Heidegger and against Heidegger, as a new way of thinking about thinking. In this view, metaphor is a movement of nearness, receiving and spontaneously creating, exteriorised in figures of speech and spatialised in material figures, and resynthesising discrete elements that emerge only after their new interactions are articulated, materialised and performed. This simultaneous emergence of discretisation and of metaphorical transformation can potentially inform any theory of difference and free some space for the consideration of the role of human interpretation. Heidegger's intuition on this connection is mostly reflected in his strong belief concerning the necessity of unfamiliarity. Unfamiliarly as disturbance of meaning displaces difference and allows room for its reconceptualisation by virtue of *hermeneia*. This displacement of difference, I argue, does not seek to replace the twofold movement of différance, but rather seeks to take into account the neglected synthesising and interpretive aspects that make any kind of differantial movement possible.

Through this new conceptualisation of metaphoricity, we can begin to think more deeply about the processes of exteriorisation, inherent in both technology and language, allowing the spatialisation of temporal processes and the temporalisation of spatial features, and indeed begin to think about the ways these metaphorical exteriorisations organise the human being. Such discussion shows that technology and language are processes of human formation, while education, namely, the one endowed with this role, needs to fight against all odds. This is because, technology is always already involved in all stages of education's unfolding and persistently reshapes the complicated nexus of connections that constitute society, education, and their interconnectedness. Technology also co-constructs, through its complicated interactions with language, the relational substrata we call home, already-there and familiarity. Reading and writing, for example, can be thought as technological and languaging modes of formation that need to be purposefully learned, applied to and, in some respect, automated, in order for education's aims to be achieved. In this respect, education's unfolding is always already entangled with technologies and with decisions about technologies.

Current digital technologies, like search engines and social networking sites, seem to be in some respect quite similar to other technologies in that they, too, like shoes and bridges, are concerned with nearness, which is in their case not a by-product of their use but the very thing they instantiate. Search engines bring close knowledge, and SNSs bring people closer to the things they care about. In another respect, and, when viewed from a different angle, digital technologies are different from previous technologies, in that they become the organa of nearness par excellence. However, in order to perform this role, they need to discretise human behaviour and to metaphorically resynthesise it in such ways as to allow a kind of unproblematic nearness to emerge. This kind of nearness appears to be making everything familiar; everyone and everything is close. It also involves users; it even incorporates them in their structure; indeed, as data for further manipulation. This nearness, however, does not necessarily make the users aware of its construction as nearness. In this respect, Heidegger was right; modern technology forces us to get into the picture, but he was also wrong in the respect that this picture is dynamic, ever-changing and possibly changeable by us.

Education's responsibility in this respect is translated first into concern for the cultivation of various modes of nearness and of creative types of connectedness that may potentially allow future citizens to create different discretisations and new schematisations of nearness. Second, this concern is also to be transfigured into the cultivation of the claim of political rights that demand access to and literate awareness of that which forms the *polis* and the political milieu. Third, education needs to view with scepticism the metaphors that construct the human-technological relations. Metaphors of the 'harnessing the educational potential' type are quite misleading in the respect that some of the new technologies are not actually new, that some technologies are not designed for learning (even though we may force them to illustrate such potential), and that these very metaphors constitute the 'idle talk' produced by designers and marketers of technology. This metaphorical banter, along with the actual ontologies created by design, attempt to sell us the supposed affordances of technology, but these affordances should be a matter of critical investigation and indeed be seen in connection to the actual modalities of human learning. Education should begin deconstructing some of these metaphors and the images that digital technologies impose; it should also establish and foster images-metaphors, which are not restrictive concerning the potential of learning and of imagination, and it should experiment with new metaphors concerning the nature of technology. Fourth, education should focus on the teaching not of selected memories in the form of subjects but of basic modes of connectedness like metaphoricity, that allows students to cultivate that which is most necessary in a world in which skills of any kind are increasingly becoming automated and distributed in unexpected exteriorisations. Lastly, as Stiegler (2009) suggests, a whole system of care should be set in place, allowing the creative and curative functioning of technologies. As he says:

> Human society is always founded on a technicity to which psychic functions are delegated and by which they become social apparatuses. The latter are supported by these *organa* that Plato called *pharmaka* – which he describes from the outset as teletechnologies, and thus as poisons. In order to imple-
> ment such remedies, while trying to avoid their functioning as poisons, a system of care is needed; this supposes, on the one hand, a pharmacopoeia, and, on the other hand, a medical science which exceeds the know-how of pharmacists.
>
> (p. 34)

Care is nothing natural: care for Heidegger is the *towards-which* that, as I have argued, is a form of synthesis, namely, a synthesis of time and of imagination that determines what comes near. For Stiegler (2010), however, it is our positioning towards care that is more important; namely our taking responsibility concerning the technology that we use and that uses us. These technologies are employed by the market, Stiegler argues, to cause a kind of '*deformation*: a destruction of the formation of the individual that education has constructed' (p. 184). It is precisely because of this phenomenon that matters of ethical responsibility arise and arise especially in connection to education and the ways in which technology can function in order to cure, that is, to form imagination without reducing it to one specific form. Ethics, in turn, takes us back to dwelling: to inhabit a world is also to bear the responsibility of thinking about this world and of those that dwell in it. It is through this consideration that ethics comes into view. Heidegger (2010) expresses this better, when he argues in the *Letter on Humanism* that:

> If the name 'ethics,' in keeping with the basic meaning of the word ēthos, should now say that 'ethics' ponders the abode of man, then that thinking which thinks the truth of Being' [ontology] as the primordial element of man, as one who ek-sists,[1] is in itself the original ethics.
>
> (p. 176)

Ethics is the thinking of the *oikos* (from the Greek word for home). This is a certain kind of economy (*oikos* + *nomos*, which means law) that legislates our comportment; it is an ecology in its original sense (*oikos* + *logos*). Our *oikos* is currently, as it has always been, a relation of nearness involving the human being and the world; it is a complicated fourfoldian relationship implicating the human being, language, technology and our house, and it is perhaps now, more than ever, a good time to start thinking about the kind of dwelling that we want to have and the kind of dwelling-beings that we aim to be.

Note

1 As pointed out, the word is written in this way in order to denote Heidegger's understanding of human temporality as a type of temporality that is continually projected into the future.

References

Heidegger, M. (2010). Letter on Humanism. In D. F. Krell (Ed.), *Basic Writings* (pp. 147–181). London and New York: Routledge Classics.

Stiegler, B. (2009). *Technics and Time 2: Disorientation* (S. Barker, Trans.). Stanford, CA: Stanford University Press.

Stiegler, B. (2010). *Taking Care of Youth and the Generations*. Stanford, CA: Stanford University Press.

Index